SYSTEMATIC BOTANY MONOGRAPHS

VOLUME 42

Systematics of Timonius Subgenus Abbottia (Rubiaceae-Guettardeae)

Steven P. Darwin

THE AMERICAN SOCIETY OF PLANT TAXONOMISTS
18 July 1994

SYSTEMATIC BOTANY MONOGRAPHS
ISSN 0373-8211

ISBN 0-912861–42–8

Printed in the United States of America

Publication of volume 42 of Systematic Botany Monographs was partly funded by Walter L. Meagher, whose generous support is gratefully acknowledged.

SYSTEMATICS OF TIMONIUS SUBGENUS ABBOTTIA (RUBIACEAE-GUETTARDEAE)

Steven P. Darwin
Department of Ecology, Evolution, and Organismal Biology
Tulane University
New Orleans, Louisiana 70118

ABSTRACT. *Timonius* subgenus *Abbottia* comprises 29 dioecious, hemiepiphytic species, including 11 varieties, distributed from eastern Borneo (Kalimantan) through Papuasia to the Philippines, tropical Australia, Fiji, and Samoa, from near sea level to 3800 m. Synapomorphies characterizing the subgenus include hemiepiphytism, truncate calyx limbs, and well-developed reticuli-paxillate leaf venation. Several characteristics of the subgenus are unique to Rubiaceae or rare in the family: the hemiepiphytic or "strangler" habit, reticuli-paxillate leaf venation formed by groups of narrow and parallel areoles, leaf blade fibers located adaxially to the vascular bundles, and leaf domatia present in abaxial axils of tertiary as well as secondary nerves. Morphological features are discussed, and keys are provided for all taxa. *Abbottia*, previously recognized as a separate genus, is here included in *Timonius* at subgeneric rank for the first time. Eight new species (*T. amungwiwanensis*, *T. bismarckensis*, *T. lamii*, *T. meridionalis*, *T. neobritannicus*, *T. oktediensis*, *T. pseudaffinis*, and *T. secundiflorus*) and four new varieties (*T. affinis* var. *samoensis*, *T. avenis* var. *magnifructus*, *T. avenis* var. *vogelkopensis*, and *T. longitubus* var. *pubescens*) are described; one new combination (*T. dolichophyllus*) is proposed.

INTRODUCTION

Ever since Valeton's pioneering work of 1909, the genus *Timonius* has been interpreted to include all dioecious species of Rubiaceae tribe Guettardeae having several to numerous ovary locules that develop into separate one-seeded pyrenes, a persistent calyx limb, and corolla lobes that are valvate or only subimbricate in bud. Roughly 180 paleotropical species conform to those broad generic criteria, making *Timonius* the largest genus of the tribe. Valeton (1909) proposed a tentative sectional framework, and Wong (1988) has suggested groupings for Malaysian species of *Timonius*, but no modern overview of the genus exists.

The present study is largely based upon an examination of all available herbarium collections of *Timonius* from Papuasia, Australia, and the South Pacific. Several weeks of field study and collecting were carried out in eastern Papua New Guinea, where material was gathered for anatomical study. Even a cursory examination of herbarium specimens revealed morphological characters that reliably define putatively monophyletic species groups within the genus; one such group has been segregated as *Timonius* subg. *Timonius* (Darwin 1993), and another of 29 species is here recognized as *Timonius* subg. *Abbottia*.

Subgenus *Abbottia* includes dioecious hemiepiphytes having reticuli-paxillate (Melville 1976) leaf venation with usually strongly reduced secondary nerves, truncate to undulate calyx limbs, and highly reduced (or absent) inflorescence bracts and bracteoles. The pistillate inflorescences are usually reduced to a solitary flower (except in *T. zuckianus* and *T. longifolius*), whereas staminate inflorescences are 3- to 25-flowered. Individuals of this subgenus are initially woody epiphytes that develop a root connection with the ground (and are then frequently described as lianas or scandent shrubs), but eventually become free-standing trees. Collectors have compared this developmental scenario with that of "strangler" species of *Ficus*, a superficial resemblance that extends likewise to the psilate leaf blades, prominent stipules, and small many-seeded fruits.

Species of subg. *Abbottia* are distributed throughout Takhtajan's (1989) Papuan Subregion with extensions to the Fijian Region as far east as Samoa (*T. affinis*), tropical Australia (*T. singularis*), the Philippines (*T. appendiculatus* and *T. pachyphyllus*), and eastern Borneo (Kalimantan, Sambaliung: *T. avenis*, *p.p.*). Thus, the western boundary of the subgenus largely conforms to Zollinger's (1857) floristic line, whereas the sole Pacific species represents yet another example of the eastward attenuation of the Malesian flora. Nineteen species of subg. *Abbottia* are endemic to New Guinea and often have relatively limited distributions there, especially if confined to higher elevations. Five species occur in the Bismarck Archipelago and Solomon Islands (Takhtajan's Bismarckian Province), one of them also in eastern New Guinea. Only *T. avenis* extends west of New Guinea, but its distribution in the Moluccas is spotty. A summary of geographical distributions is given in Fig. 1.

All but two species of subg. *Abbottia* appear to be restricted to primary or old secondary forests at elevations from near sea level to 3800 meters (Fig. 2); at high elevations, species occur in stunted scrub forests on ridgetops. Available notes suggest that *T. longifolius* (Sepik region of New Guinea) may be confined to river banks and stream margins but is probably not a true rheophyte (van Steenis 1981); *T. pubistipulus*, distributed over a wide altitudinal range, is common in successional habitats, such as forest regrowth in burned-over areas, old landslips, and trailsides.

Twenty-nine taxa are here accorded species rank on the basis of morphological discontinuities, and an attempt is made to describe infraspecific and usually geographically significant variability through recognition of a few named varieties. However, the limitations of available herbarium material precluded a detailed analysis of morphological diversity at the populational or even regional level. Furthermore, some taxa are as yet known only from either pistillate or staminate specimens, the latter with fewer distinguishing characters and thus more difficult to interpret.

TAXONOMIC HISTORY

The genus *Abbottia* was established by Ferdinand von Mueller in 1875 to accommodate the species *A. singularis*, from the vicinity of Rockingham Bay, Queensland, Australia. The generic name honors Francis Abbott (1834–1903), director of the Hobart Botanical Gardens, Tasmania. Mueller tentatively placed the genus in Rubiaceae tribe Mussaendeae [=Isertieae]. Interestingly, he did not associate *Abbottia* with the *Guettarda polyphragmoides* he described in 1861, and which Bentham (1867) had correctly recognized as a synonym of *Timonius rumphii* DC. [=*T. timon* (Sprengel) Merrill]. Nor did Mueller compare his new genus and species with the closely similar *Timonius affinis*, which Asa Gray had described from Fiji in 1858; indeed, Gray's discussion of *Timonius*, *Bobea* Gaud., and related taxa

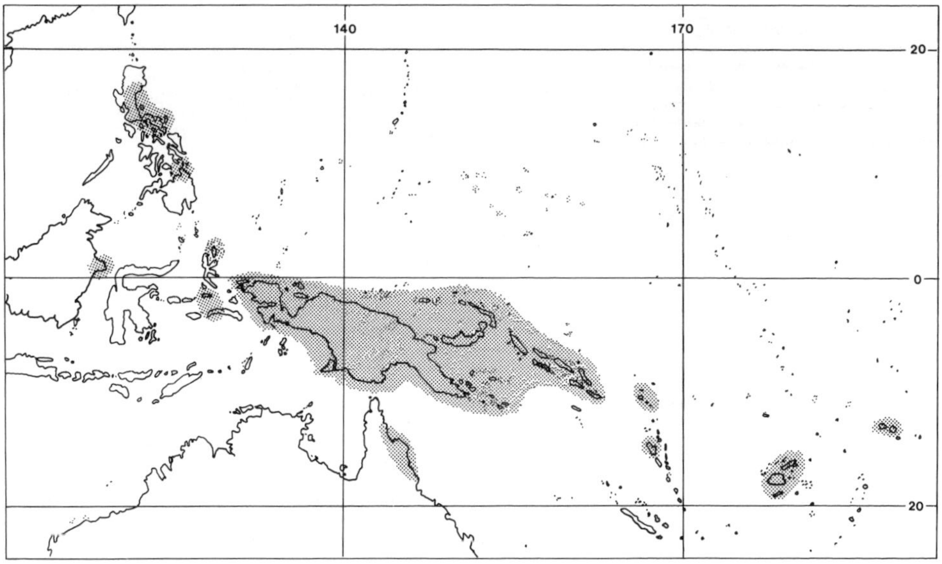

FIG. 1. Distribution of *Timonius* subg. *Abbottia*.

established the generic concepts applied to Old World Guettardeae for more than a century thereafter.

Mueller's concept of *Abbottia* and its tribal placement were followed by Bailey (1900). Schumann (1891) also maintained *Abbottia* as distinct from *Timonius*, but moved it to tribe Guettardeae, following Baillon's suggestion of a close relationship between *Abbottia* and *Timonius* (1880, p. 425, footnote). In transferring *Abbottia singularis* to *Timonius*, L. S. Smith (1957) stated that "there can be little doubt that the species belongs in the *T. avenis* group, of which several species occur in New Guinea."

A broad generic definition was taken up by Valeton in his seminal study of *Timonius* in the Netherlands Indies (Valeton 1909), but among the 33 species treated by him, only *T. avenis* Valeton is here assigned to subg. *Abbottia*. In Valeton's tentative, infrageneric classification, *T. avenis* is placed in *Timonius* sect. *Helospora* with *T. flavescens* (Jack) Baker, *T. wallichianus* (Korth.) Valeton, and fourteen other miscellaneous species that are manifestly of no very close relationship. Additional species of *Timonius* subsequently described from the Philippines by Merrill (*T. appendiculatus*, 1909; *T. pachyphyllus*, 1925), from New Guinea by Wernham (*T. carstensensis*, *T. heptamerus*, and *T. wollastonii*, 1916), by Moore (*T. cryptophlebus*, 1923), and again by Valeton (*T. longifolius*, 1927), are all here placed in subg. *Abbottia*.

Valeton's inclusive concept of *Timonius* was accepted by Merrill and Perry (1945) in their evaluation of Archbold Expedition collections from Papuasia. They recognized eight species (including four new ones) as being of the *T. avenis* alliance, but did not specify the characters by which the group could be recognized; two others of their new species, *T. longitubus* and *T. bougainvillensis* were said to be of uncertain relationship but are here also placed in subg. *Abbottia*.

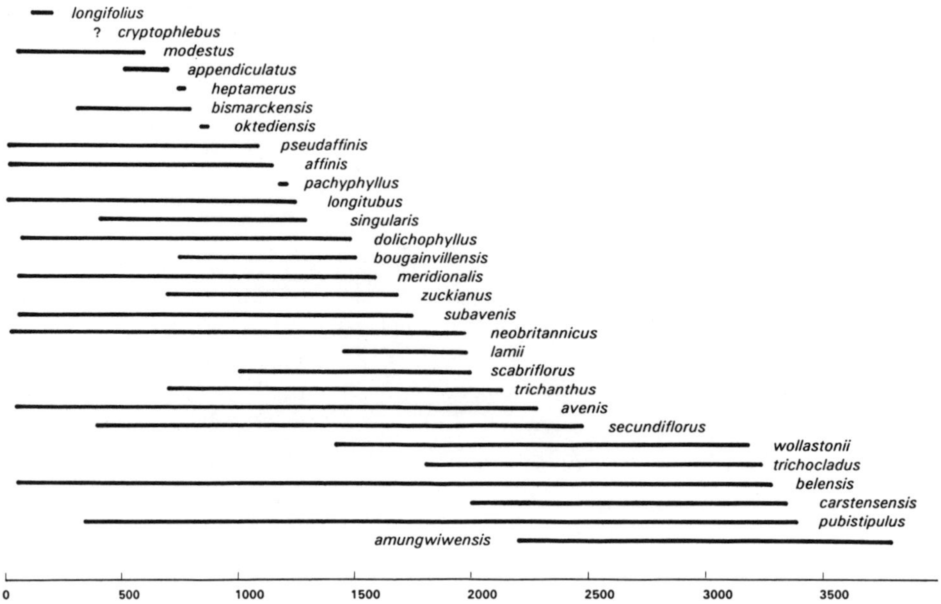

FIG. 2. Altitudinal distribution of species of *Timonius* subg. *Abbottia*, based on data from herbarium collections. Scale in meters.

MATERIALS AND METHODS

This study is based on available herbarium collections (roughly 500 "numbers") borrowed from 34 institutions as well as field work in Papua New Guinea. FAA-fixed material of *T. pubistipulus* var. *pubistipulus* (*LAE 77669* and *77670*) was sectioned for anatomical study using standard dehydration and paraffin methods, or was hand-sectioned and stained. Trichomes and crystals were studied under a light microscope equipped with polarizing filters. Leaf venation was studied from herbarium material cleared according to the method of O'Brien and Teichman (1974) and photographed under polarized light. Pollen grains from herbarium specimens were acetolyzed [modification of Erdtman's (1960) method] and then mounted in glycerin jelly for examination under the light microscope. Attempts to locate and count chromosomes of pollen sporocytes in field-collected material were not successful; no chromosome count for any species of subg. *Abbottia* has yet been reported.

MORPHOLOGY

Habit. Species of subg. *Abbottia* have been recorded as epiphytic shrubs, lianas, climbers, stranglers, or large forest trees. Judging from available specimen data, most if not all species are hemiepiphytes that eventually become free-standing trees through production of woody, adventitious roots. At least ten species eventually become trees more than 20 m tall, and some (e.g., *T. avenis*) reach 40 m and 80 cm dbh. As previously stated, collectors frequently note a similarity to "strangler" species of *Ficus*, a resemblance reinforced by the conspicuous stipules and the several- to many-seeded, fleshy fruits. For a few species, epiphytism has not been reported, but in those instances habit data are either entirely missing from specimens, or the available herbarium collections are few in number; one may assume that plants in an early, epiphytic stage of growth would be less frequently collected than older, larger, terrestrial individuals. Nonetheless, it now appears that subg. *Abbottia* is the largest group of hemiepiphytes known in the Rubiaceae, a growth habit that is otherwise extremely rare in the family (Robbrecht 1988).

Bark and wood. The bark of mature individuals (trees or shrubs) is reported as smooth and pustular or scaly to fissured, varying in color from gray to brown; lenticels are often prominent. The wood is hard, white to straw-colored or rarely reported as dark red (*T. avenis* var. *pubipetalus*). Ter Welle et al. (1983), reporting on wood anatomy in Guettardeae, concluded that *Timonius* is anomalous in the tribe; however, no specimen of subg. *Abbottia* was represented among the material they examined. Whether the anatomy of wood taken from adventitious roots (the "trunks" of mature plants) differs significantly from that of true branches is as yet unknown.

Trichomes and crystals. Pubescence on vegetative and reproductive parts is highly variable, but usually constituted of septate hairs with thick, striate cell walls. Shorter, smooth-walled hairs may be intermixed, and these have rhomboidal crystals embedded in their walls, as is characteristic of tribe Guettardeae (Verdcourt 1958; Robbrecht 1988). The crystal-containing hairs are usually found on outer surfaces of stipules and corollas. Secretory colleters of the "standard" conical type (Robbrecht 1988) are present near interior bases of stipules and sometimes also distributed along the stipule midrib. Similar colleters are also occasionally present at the interior base of the free calyx tube.

In addition to the rhomboidal trichome crystals, cells containing crystal sand are common and scattered throughout primary plant tissues, especially ovary and fruit wall. Raphid crystals were not seen.

Leaves. The leaves are opposite, decussate, petiolate, and isophyllous with usually elliptic or rhombic, ovate, or obovate blades (1–) 3–16 (–20) cm long, 1–6 (–9) cm wide (Fig. 3). In most species of the subgenus, the leaf blades lack prominent lateral nerves, the degree of lateral nerve reduction roughly correlated with blade thickness: in thin-textured blades (e.g., *T. bougainvillensis*), the lateral nerves are prominent, whereas in thick-coriaceous blades (as in species from high elevations, e.g., *T. wollastonii*), the lateral nerves are indistinguishable from higher-order venation. The fine venation is, in the terminology of Melville (1976), reticuli-paxillate, characterized by groups of parallel, elongate areoles without free-ending veinlets (Fig. 4). In species of *Timonius* other than of subg. *Abbottia*, the venation is highly variable but not reticuli-paxillate; in subg. *Timonius*, for example, the veinlets are dichotomo-reticulate to dendro-reticulate (Darwin 1993). A detailed survey of venation patterns in tribe Guettardeae has not been conducted, but venation patterns superficially similar to those of subg. *Abbottia* have been reported in *Antirhea* Comm. ex Juss. (Chaw & Darwin 1992) and *Neolaugeria* Nicolson (Nicolson 1979), as well as other genera of Rubiaceae and some Ochnaceae (Nicolson 1979) and Quiinaceae (Foster 1950, 1952). In subg. *Abbottia*, the leaf venation pattern is accen-

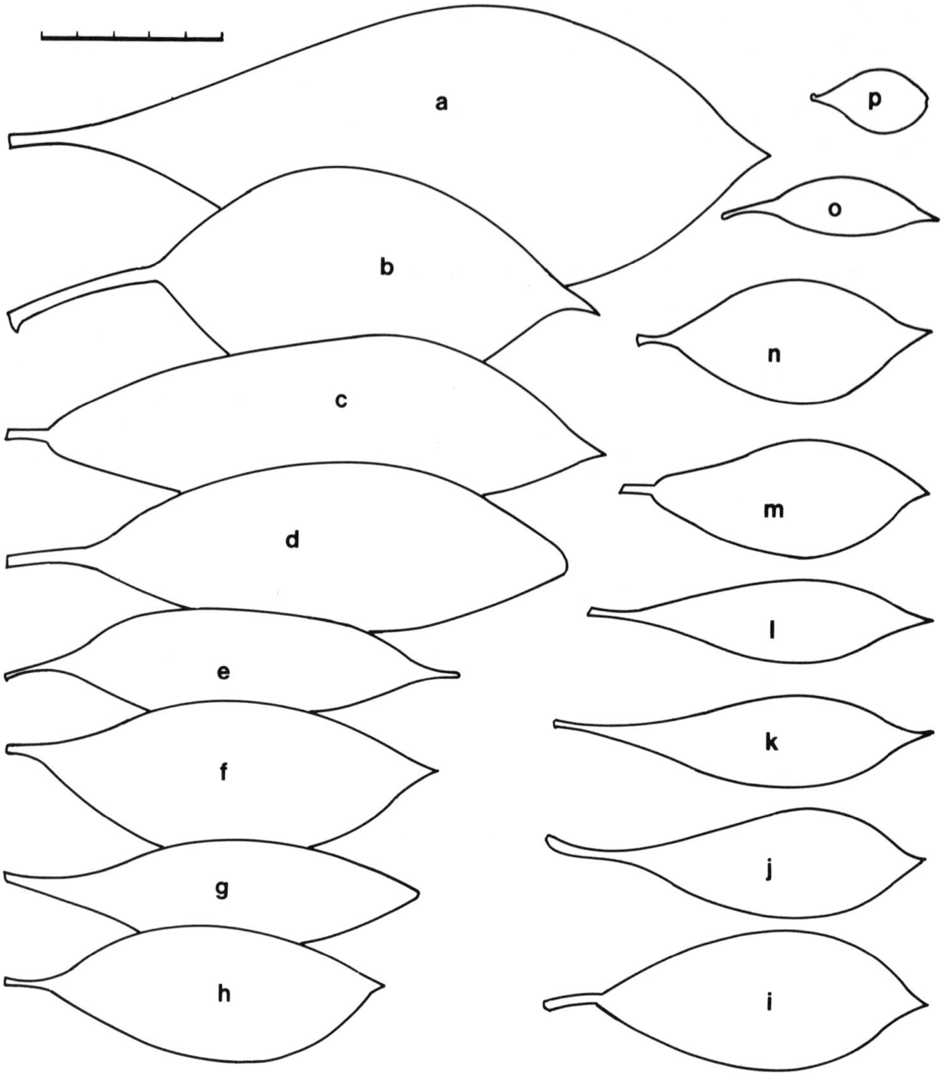

FIG. 3. Outlines of representative mature leaves from selected species of *Timonius* subg. *Abbottia*. Consult species descriptions for extent of variability in leaf blade size and shape. a. *T. longitubus* (*Kajewski 2009*). b. *T. zuckianus* (*Brass 28458*). c. *T. longifolius* (*Darbyshire & Hoogland 8355*). d. *T. pachyphyllus* (*Bur. Sci. 40794*). e. *T. bougainvillensis* (*NGF 31382*). f. *T. bismarckensis* (*LAE 65579*). g. *T. appendiculatus* (*Philippine Nat. Herb. 10583*). h. *T. trichocladus* (*Brass & Versteegh 10486*). i. *T. trichanthus* (*Brass 32245*). j. *T. neobritannicus* (*LAE 76014*). k. *T. pseudaffinis* (*Roy. Soc. Exped. 2876*). l. *T. subavenis* (*LAE 73981*). m. *T. oktediensis* (*NGF 42793*). n. *T. meridionalis* (*Brass 27532*). o. *T. avenis* var. *avenis* (*BW 10275*). p. *T. wollastonii* (*Lam 1485*). Bar = 5 cm.

tuated by a reticulum of elongate, fusiform fibers with thick, striate, nonlignified walls. The fibers lie parallel with the veins and veinlets, and are "stacked" in uniseriate (sometimes biseriate) files that extend adaxially from them (Fig. 5). In one specimen studied in detail (*T. pubistipulus* var. *pubescens*), fibers were found to be associated with all vascular bundles of the leaf except the costa; the uniseriate files

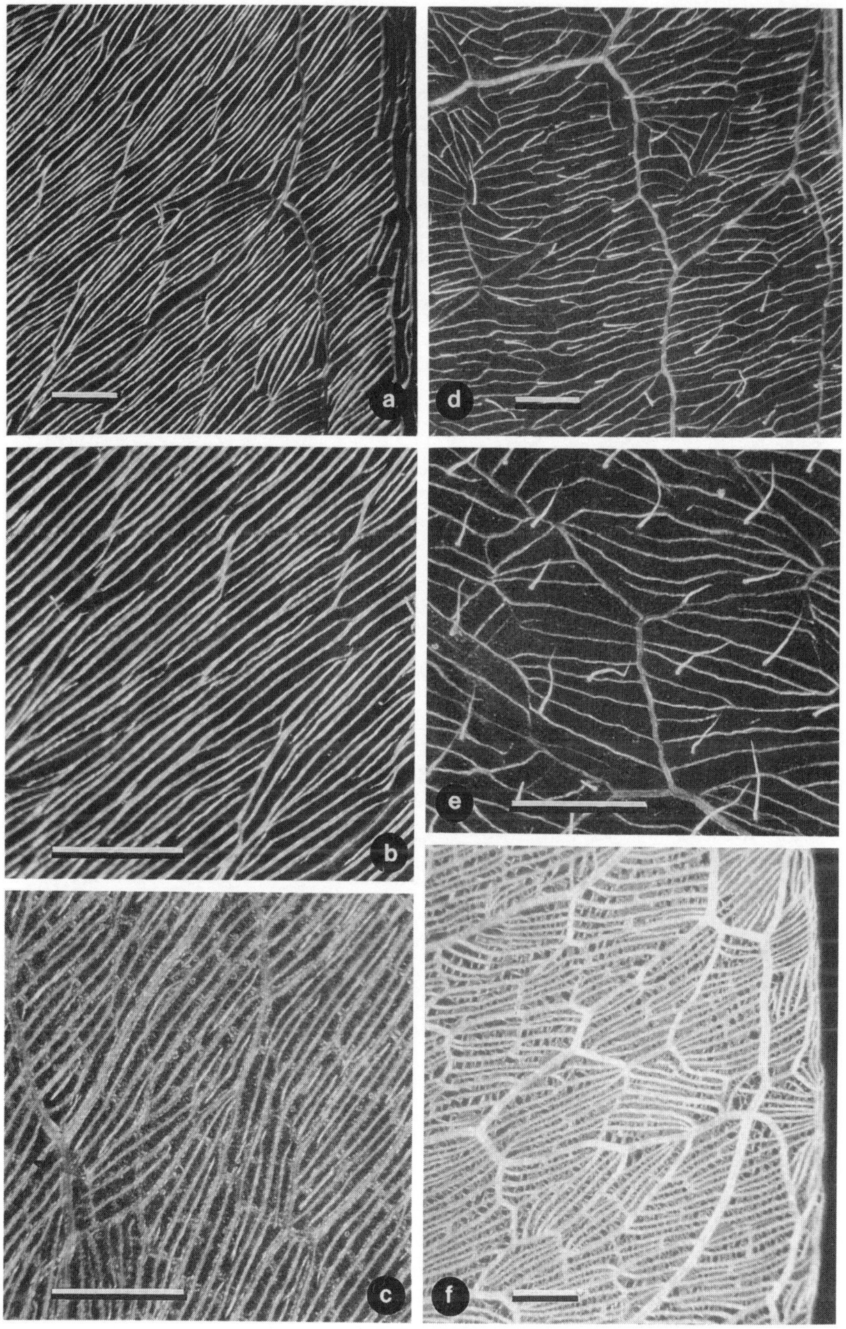

FIG. 4. Cleared leaf blades of species of *Timonius* subg. *Abbottia* photographed under polarized light. a, b. *T. neobritannicus* (*NGF 21945*). c. *T. secundiflorus* (*Brass 13310*). d, e. *T. longifolius* (*Darbyshire & Hoogland 8355*). f. *T. carstensensis* (*Brass 4705*). Bar = 1 mm.

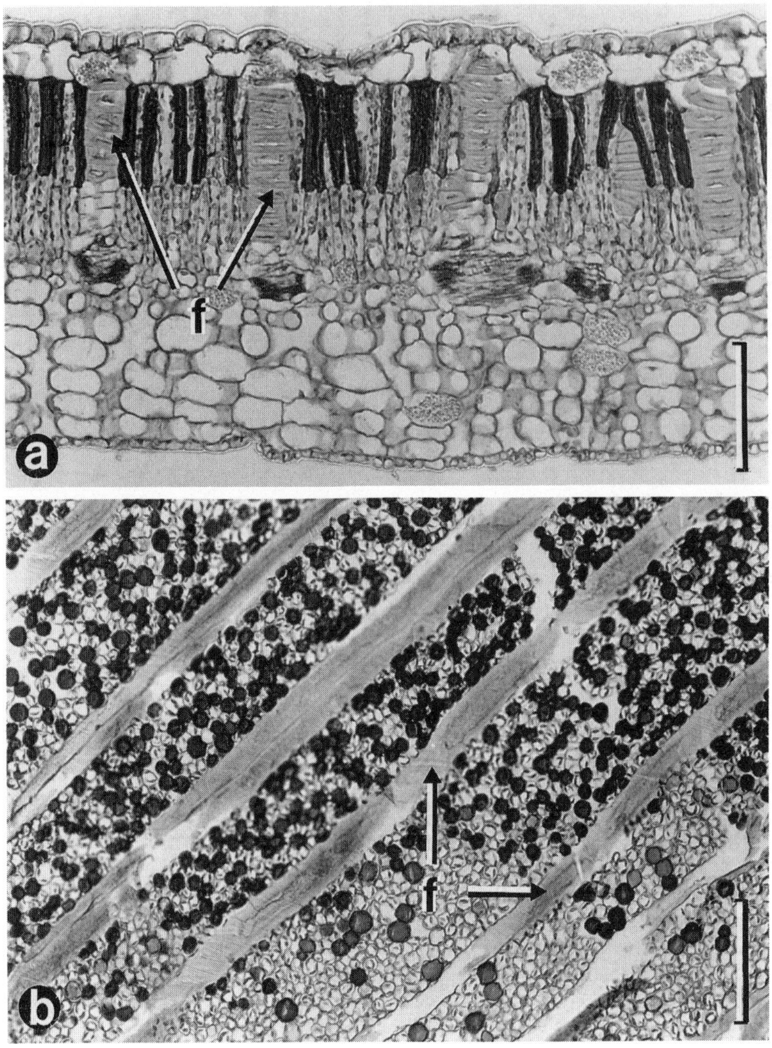

FIG. 5. Leaf anatomy of *Timonius pubistipulus* var. *pubescens* (*LAE 77669*). a. Cross section of blade showing vascular bundles and associated abaxial fiber cells (f). b. Paradermal section of blade showing palisade parenchyma and fibers (f). Bars = 100 μm.

were as much as 15 cells tall, extending as far as the upper leaf hypodermis (Fig. 5). Fibers (or fiber-tracheids) are associated with the leaf venation in several rubiaceous genera (Metcalfe & Chalk 1950, 1979), but have been described as forming a superficial network on the *abaxial* surface of the leaf. Adaxial fibers, as reported here, have been seen in *Antirhea* (Chaw & Darwin 1992) and are likely to be found in other genera of Guettardeae.

Leaf domatia. Domatia are often present on abaxial blade surfaces and take the form of open pits, shallow to deep pockets, or cryptlike chambers (for terminology, see Jacobs, 1966). The domatia are almost always situated in the axils of secondary nerves, but pit-domatia are also often positioned over tertiary nerve axils, and with reduction of the venation, they appear to be scattered over the blade. As far as I am aware, association of domatia with tertiary nerves is a feature not previously reported for Rubiaceae. Deeper, partially covered pocket-domatia are usually associated with the costa, in the axils of secondary nerves if the venation is at all well developed; the abaxial leaf epidermis frequently covers a large interior chamber except for a small ostiole or slitlike opening in the domatium "roof." A pair of large pocketlike or cryptlike domatia flank the base of the costa in several species of subg. *Abbottia*. Domatia are glabrous to densely pubescent. The domatioid extrafloral nectaries found in some species of subg. *Timonius* (Darwin 1993) were not seen in collections of subg. *Abbottia* examined for the present study.

Stipules. The interpetiolar stipules are mostly lanceolate, acuminate, and individually caducous. In terminal buds, the stipules are dextrorsely or sinistrorsely imbricate, sometimes in both directions on the same specimen. The stipules are individually deciduous and thus very different from the calyptrate stipular sheaths characteristic of subg. *Timonius*, but similar to stipules found in many other species of *Timonius* (e.g., *T. laugerioides* Wernham) and of the type probably ancestral for the genus. In some species of subg. *Abbottia*, stipules regularly exceed 10 cm in length (13 cm the longest seen in *T. zuckianus*). Abaxial and adaxial pubescence is highly variable and often a reliable species character; in all specimens examined, the adaxial stipule bases are invested with long, secretory colleters, whether or not other indument is present.

Inflorescences. All species of subg. *Abbottia* were found to be strictly dioecious (staminate and pistillate structures on different plants). The pedunculate cymes are always solitary in distal leaf axils; staminate and pistillate inflorescences are usually dissimilar in size and number of flowers. Staminate inflorescences are generally 2-armed and 3- to 25-flowered; the peduncles are 1–14 mm long; bracts and bracteoles are absent. Pistillate inflorescences are usually 1-flowered, but 2- or 3-flowered specimens were found in *T. zuckianus* and 5-flowered inflorescences were occasionally seen in *T. longifolius* (Fig. 15d); peduncles reach a length of 25 mm in fruit; bracts and bracteoles are usually absent or much reduced.

Calyx. The staminate calyx is cupular to subcylindrical, usually not more than 3 mm long (measured from very base) but reaching 10 mm in *T. longitubus*. The pistillate calyx is similar in shape, usually about 2 mm long (measured from top of ovary) but to 6 mm in *T. longitubus* (Fig. 15b, c). In all species, the calyx tube is truncate to irregularly and shallowly undulate or rarely dentate, sometimes irregularly rupturing with expansion of the corolla.

Corolla. Corollas of subg. *Abbottia* show relatively little variability except in dimensions and exterior pubescence, although the appendaged corolla lobes of the two Philippine species (*T. appendiculatus* and *T. pachyphyllus*) are distinctive: three appendages arise from the interior apex of each reflexed corolla lobe to form a brushlike limb (Fig. 11c). Staminate corollas of all species are slenderly infundibular with 4 (–6) lobes and tubes 8–22 mm long. Pistillate corollas are often somewhat broader and shorter than the staminate and (5-) 6- to 8- (12-)lobed; the corolla tubes are 4–15 mm long. Corollas of staminate and pistillate flowers vary from glabrous to densely strigillose outside, but are uniformly glabrous within. Corolla lobe aestivation (Fig. 20h, i) is of an interlocking type, well described by Valeton (1909), and here considered diagnostic for *Timonius* (Darwin 1993).

Mature pistillate corollas are not abundant in herbarium collections and probably last only one day. In field notes, corollas are described as white or yellowish to pale green and often flushed with pink outside; corolla interiors and lobes are paler to white. Pollinators have not been identified, and the flowers are said to be odorless or only faintly scented. In most well-sampled species, flowers have been collected throughout the year, but reproduction may be seasonal in *T. avenis*, *T. bougainvillensis*, and perhaps other species.

Androecium and pollen. In staminate flowers, the usually 4 stamens are dorsally attached by a short filament to the corolla tube well below the throat and included within the corolla except for the sometimes protruding anther tips. The anthers are 3–10 mm long, glabrous, linear, usually somewhat sagittate at base where the locule pairs are separate. Transverse septa were not seen. In pistillate flowers, the androecium is represented by sterile staminodia that are usually attached at or below the middle of the corolla tube. The abortive anthers are 2–3 (–5) mm long, glabrous, number as many as the corolla lobes and alternate with them; anther chambers may be present, but no pollen was observed in any of them.

Pollen of subg. *Abbottia* was investigated from 39 herbarium specimens representing 18 species; after acetolysis, grains were examined under the light microscope. Pollen of subg. *Abbottia* is isopolar, semitectate, radiosymmetrical, and subspheroidal. The grains frequently appear to be inaperturate to vaguely or strongly 3-colpate, and there is often some variability in aperture development among grains from the same anther locule. Inaperturate grains have been reported for *T. appendiculatus* (Luo et al., 1993), whereas invariably colpate pollen has been found in species of *Timonius* from Borneo (C. Puff, pers. comm.). Similar variability has also been reported in species of subg. *Timonius* (Darwin 1993) and may represent different stages of pollen development. In general morphology, pollen of *Timonius* is much like that of *Antirhea* (Chaw & Darwin 1992) and *Bobea* (Darwin & Chaw 1990).

In specimens examined, size (longest diameter) of pollen grains ranged from 21 to 44 μm (rarely to almost 53 μm) (Table 1). In view of the limited sample sizes and absence of karyological data, average pollen diameter was not used as a taxonomic character, since grain size may be correlated with ploidy-level (Puff 1986).

Gynoecium. The inferior ovary (hypanthium of pistillate flowers) is ovoid, subglobose, or obovoid, and surmounted by a glabrous disc, this sometimes expanding in fruit (especially in *T. affinis* and *T. lamii*). The terminal and glabrous style is (3–) 5–18 mm long, and is distally divided into (4–) 6–8 (–12) unequal, exserted stigmatic arms; the number of arms usually equals the number of corolla lobes. The interior of the style is occupied by a solid, vertically ribbed column of stigmatoid (pollen-tube-conducting) tissue composed of elongate parenchyma cells with dense cytoplasm; the number of ribs equals the number of style arms. At the distal end of the ovary, the stigmatoid tissue ramifies into thinner strands that each terminate at the apex of an ovary locule (Darwin 1993). The ovary interior is occupied by 12 to 80 (or more?) 1-seeded locules that radiate in double-files from the apical region of the ovary, or the locules are essentially vertical and more-or-less parallel and the ovules pendulous. There are as many double-files of locules as there are style branches.

The structure of the ovary of *Timonius* has been studied in some detail by Martinello (1992), who interpreted the gynoecium as representing an increase in the number of carpels (locules) caused by secondary formation of successive whorls. Puff (pers. comm.) considers this comparable to the situation in *Punica* or the navel orange and perhaps unique in Rubiaceae. Within each locule, the solitary ovule is attached to an apical placenta (Martinello 1992; not the "placenta" as defined by Darwin, 1993); during the course of development, the wall of each locule becomes a fibrous, lignified endocarp (pyrene), and the placenta becomes a plug or "stopper" of lignified stone cells that occludes an opening at the apical (adaxial) end of the pyrene. Except for the less extensively developed stigmatoid tissue and usually fewer ovules, ovary structure in subg. *Abbottia* is much like that of subg. *Timonius* (Darwin 1993).

In staminate flowers, the gynoecium is represented by a slender, unbranched, glabrous pistillodium 2–5 (–8) mm long. This structure was found to be of solid tissue (thick-walled parenchyma with scattered tannin-filled and crystaliferous cells); no chambers were observed in specimens sectioned, although small, empty locules have been found in pistillodia of other species of *Timonius* (Darwin 1993).

Fruits and seeds. The drupaceous fruits are obovoid to subglobose or short-cylindrical, mostly 5–18 (–25) mm long and wide, thinly fleshy, dark red to purple or black when mature, and surmounted by the disc and persistent calyx limb. Field notes report them eaten in quantity by forest birds. The pyrenes number 15–350 (–500) per fruit and are variously oriented (Figs. 6, 7); in the majority of species, pyrenes radiate outward and downward from the distal end of the ovary ("oblique" pyrene orientation), whereas in others (e.g., *T. avenis*) the pyrenes are vertical or subvertical and parallel.

In three species from the eastern end of the subgeneric range (*T. affinis*, *T. bougainvillensis*, and *T. longitubus*), the mature fruits develop a central, schizogenous cavity (Fig. 7d–h). This feature was also noted in subg. *Timonius* (Darwin 1993), where it is believed to be an aid in water-dispersal and functionally equivalent to the "floating organ" found in fruits of *Guettarda speciosa* L. (Robbrecht 1988, p. 129). However, similar but smaller schizogenous cavities sometimes form in fruits of *T. belensis* from upland New Guinea, and a lack of accurate field data leaves open the possibility that the chambers are artifacts of specimen drying.

The small pyrenes are ovoid to oblong and often variously compressed. The wall is thickest toward the abaxial end, becoming fimbriate at the adaxial (placental) end where a germination opening is, as stated, occluded by a plug of thick-walled, lignified stone cells, these representing a metamorphosed placenta (Martinello 1992). The elongate, torpedo-shaped embryo with two small cotyledons is surrounded by a thin endosperm. Walls of outermost testa cells are thin and appear to be unornamented.

Table 1. Pollen grain sizes (μm) for selected collections of *Timonius* subg. *Abbottia;* minimum, maximum, and average diameters seen in a sample of fifteen acetolyzed grains.

	Minimum	Maximum	Average
T. amungwiwanensis			
NGF 36847	33.4	52.8	35.9
T. avenis var. avenis			
Kostermans 5229	27.3	34.3	30.4
T. avenis var. pubipetalus			
LAE 59583	28.6	39.6	33.4
Brass 12878	25.5	35.2	29.9
T. avenis var. vogelkopensis			
Kanehira & Hatusima 13141	22.9	30.8	27.8
BW 15353	26.4	30.8	29.4
T. belensis			
Pullen 5216	28.2	35.2	31.9
T. dolichophyllus			
NGF 33179	31.7	40.5	37.5
NGF 33179	30.8	44.0	36.8
T. lamii			
BW 11462	25.5	35.2	27.7
Lam 1572	25.5	30.8	27.9
T. longifolius			
Hoogland & Craven 10747	22.9	30.8	25.5
T. longitubus var. longitubus			
Brass 3244	27.3	37.8	33.0
T. meridionalis			
Brass 24796	26.4	33.4	30.3
Brass 28630	26.4	31.7	29.7
T. modestus			
Brass 7076	27.3	32.6	34.6
Jacobs 9230	21.1	27.3	25.0
T. neobritannicus			
NGF 27241	26.7	30.8	28.3
NGF 21945	22.0	27.3	24.9
T. oktediensis			
NGF 42793	22.9	30.8	26.9
T. pseudaffinis			
Roy. Soc. Exped. 6319	25.5	31.7	27.8
T. pubistipulus var. pubistipulus			
NGF 32181	29.0	40.5	36.0
Pullen 7842	29.9	39.6	34.5

Table 1 continued.

	Minimum	Maximum	Average
T. pubistipulus var. pubescens			
Hartley 11725	26.4	31.7	28.9
T. scabriflorus			
Brass 32230	24.6	29.0	26.6
Clemens 3586A	26.4	30.8	28.8
T. secundiflorus			
Hoogland & Craven 10950	22.9	27.3	25.7
Ledermann 8479	22.0	26.4	23.7
Ledermann 8479	26.4	32.6	29.7
T. subavenis			
Ledermann 8793	26.4	31.7	29.7
T. trichanthus			
NGF 4950	36.1	40.5	38.9
Hartley 12078	29.9	34.3	32.2
T. trichocladus			
Brass 9504	26.4	34.3	30.9
Brass 9504	27.3	35.2	31.0
T. wollastonii			
Römer 1242	31.7	25.5	28.3
T. zuckianus			
Paijmans 156	26.4	35.2	29.9

PHYLOGENETIC CONSIDERATIONS

Timonius subg. *Abbottia* is here considered to be a monophyletic taxon whose member species share a hemiepiphytic habit, reticuli-paxillate leaf venation, and truncate calyx limbs; these are interpreted as representing synapomorphies—derived homologous similarity due to common ancestry. All of those character states are known to occur in species of *Timonius* not placed in subg. *Abbottia* (e.g., epiphytes in *T. epiphyticus* Elmer) or even in other genera of Guettardeae (e.g., reticuli-paxillate venation in *Neolaugeria* of the West Indies), but such occurrences are here interpreted as homoplasies.

Other features shared among species of subg. *Abbottia*, such as separate pyrenes, separate and imbricate stipules, and dioecism, are characteristic of *Timonius* as a whole, and thus are probably symplesiomorphies in subg. *Abbottia*. The unusual corolla aestivation of tightly interlocking lobes is very likely a synapomorphic feature shared by all species of *Timonius*, and is the principal morphological evidence that supports the recognition of *Abbottia* at subgeneric rank rather than as a segregate genus.

Firm hypotheses as to phylogenetic relationship between subg. *Abbottia* and other species of *Timonius* or species-groups must await a survey of the genus throughout its geographical range, although it is here conjectured that such species

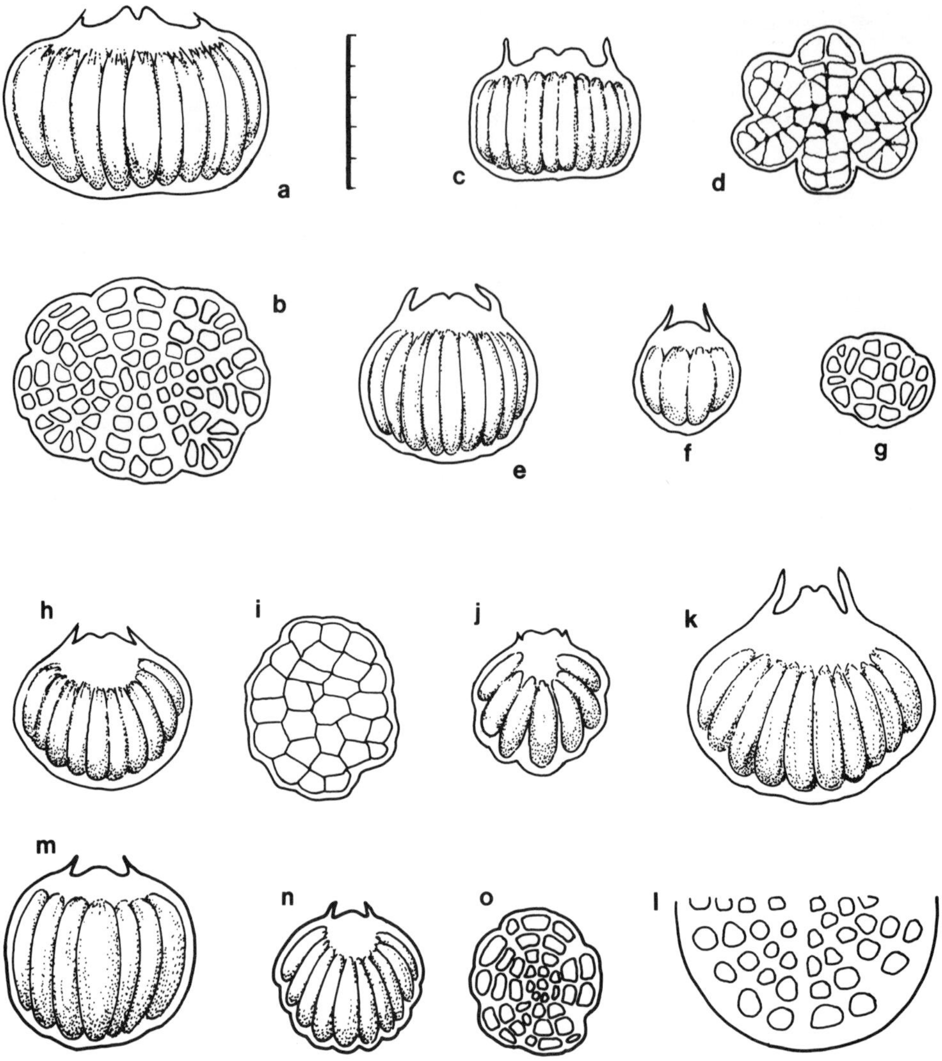

FIG. 6. Sections (semidiagramatic) through fruits of selected species of *Timonius* subg. *Abbottia*. a. *T. avenis* var. *magnifructus* (*Kostermans 2220*), longitudinal section. b. Same, cross section. c. *T. longifolius* (*Hoogland & Craven 10663*), longitudinal section. d. Same, cross section. e. *T. trichanthus* (*ANU 2719*), longitudinal section. f. *T. wollastonii* (*Lam 1485*), longitudinal section. g. Same, cross section. h. *T. singularis* (*L. S. Smith 3755*), longitudinal section. i. Same, cross section. j. *T. pseudaffinis* (*BSIP 15191*), longitudinal section. k. *T. secundiflorus* (*Brass 13310*), longitudinal section. l. Same, cross section. m. *T. pubistipulus* var. *pubistipulus* (*Carr 13819*), longitudinal section. n. *T. appendiculatus* (*Elmer 16636*), longitudinal section. o. Same, cross section. Bar = 5 mm.

as *T. laugerioides* have retained several plesiomorphic (ancestral) morphological character states, such as isomorphic staminate and pistillate inflorescences, unspecialized leaf venation, and a terrestrial habit. Unfortunately, too few morphological characters are available to support a detailed cladistic analysis of subg. *Abbottia*, especially because staminate or pistillate collections are still unknown for some species. In many instances, too few collections are available to evaluate accurately

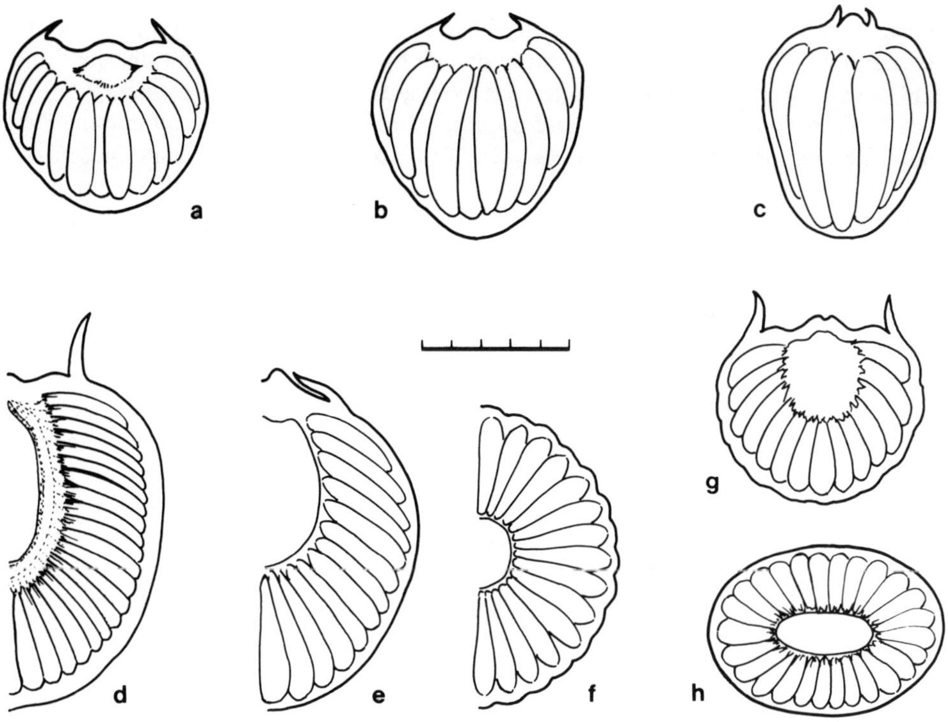

FIG. 7. Sections (semidiagramatic) through fruits of selected species of *Timonius* subg. *Abbottia*. a. *T. belensis* (*Brass 31093*), longitudinal section. b. *T. amungwiwanensis* (*Moi 107*), longitudinal section. c. *T. trichocladus* (*Brass & Versteegh 11102*), longitudinal section. d. *T. longitubus* (*Kajewski 2009*), longitudinal section. e. *T. affinis* (*A. C. Smith 5507*), longitudinal section. f. Same, cross section. g. *T. bougainvillensis* (*Schodde & Craven 3911*), longitudinal section. h. Same, cross section. Bar = 5 mm.

character variability within species, especially for such features as leaf blade shape, indument, dimensions of flower parts, and number of ovary locules. Nevertheless, several groups of species may be discerned on the basis of morphological similarities, and some of those groups probably represent monophyletic lineages.

Timonius appendiculatus and *T. pachyphyllus* are the only species of subg. *Abbottia* known from the Philippines. Both taxa have three prominent, fingerlike projections on the inner tips of the corolla lobes (Fig. 11c), and these appear not to be the result of compression during aestivation. Most other species of *Timonius* have variously ridged corolla lobe surfaces, but this condition results from the peculiar corolla aestivation and never produces the "calli" seen in the Philippine taxa.

Timonius avenis, *T. cryptophlebus*, *T. dolichophyllus*, *T. lamii*, *T. longifolius*, *T. subavenis*, *T. trichanthus*, *T. wollastonii*, and *T. zuckianus* all have pyrenes that are oriented vertically or subvertically, the mature fruits appearing depressed-globose to depressed-cylindric. In *T. carstensensis* and *T. heptamerus*, for which mature fruits were not available for study, a vertical orientation of mature pyrenes may be inferred from the arrangement of ovary locules. Species of this group extend from the western end of the subgeneric range (*T. avenis* in Borneo) as far east as southeastern Papua (absent from the Philippines and Australia), and in elevation from near sea level to well above 3000 meters (*T. carstensensis* and *T. wollastonii*). Except for *T. avenis*, these species have relatively narrow distributions and perhaps were independently derived from a more widespread ancestor.

In *T. amungwiwanensis*, *T. pubistipulus*, and *T. trichocladus*, all confined to New Guinea, the peripheral pyrenes of the fruit are usually shorter than the central ones, the fruits thus becoming obovoid when mature. This pyrene arrangement is distinctive and not intermediate between the vertical and oblique arrangements found in other species of the subgenus.

Twelve species (*T. affinis*, *T. pseudaffinis*, *T. belensis*, *T. bismarckensis*, *T. bougainvillensis*, *T. longitubus*, *T. meridionalis*, *T. modestus*, *T. neobritannicus*, *T. scabriflorus*, *T. secundiflorus*, and *T. singularis*) have subequal pyrenes that radiate from near the apex of the fruit, and thus comprise a vaguely discernible species group. Pistillate inflorescences are always 1-flowered (versus 2- to 9-flowered inflorescences in some other species of the subgenus), but solitary pistillate flowers are found in many other species of *Timonius* outside subg. *Abbottia*, and may be an ancestral characteristic for the subgenus. Likewise, the prominent secondary nerves in leaf blades of *T. bougainvillensis* may be plesiomorphic for the subgenus; other species show different degrees of nerve reduction.

One remaining species, *T. oktediensis* from south-central New Guinea, is placed in subg. *Abbottia* on the basis of its epiphytic habit and truncate calyx limb, but in the absence of pistillate specimens its relationship to other species of the subgenus remains in doubt.

TAXONOMY

Timonius DC., Prodr. 4: 461. 1830, nom. cons.—TYPE: *Timonius rumphii* DC., nom. illeg. [≡*Timonius timon* (Sprengel) Merrill ≡ *Erithalis timon* Sprengel, typ. cons.].

Porocarpus Gaertner, Fruct. sem. pl. 2: 473. 1791, nom. rejic.—TYPE: *Porocarpus helminthotheca* Gaertner.

Polyphragmon Desfontaines, Mém. Mus. Hist. Nat. Paris 6: 5. 1820, nom. rejic.—TYPE: *Polyphragmon sericeum* Desfontaines.

Helospora Jack, Trans. Linn. Soc. London 14: 127. 1823, nom. rejic.—TYPE: *Helospora flavescens* Jack.

Burneya Chamisso & Schlechtendal, Linnaea 4: 188. 1829, nom. rejic.—TYPE: *Burneya forsteri* Chamisso & Schlechtendal, nom. illeg. [≡*Erithalis polygama* Forster f.].

Eupyrena Wight & Arnott, Prodr. fl. Ind. orient. 422. 1834.—TYPE: *Eupyrena glabra* Wight & Arnott.

Abbottia F. v. Mueller, Fragm. 9: 181. 1875, non *Abbotia* Rafinesque (Juncaginaceae), 1836.—TYPE: *Abbottia singularis* F. v. Mueller.

Dioecious or rarely polygamodioecious trees, shrubs, or woody epiphytes. Stipules prominent in terminal buds, valvate, imbricate, or calyptrate. Leaves simple, opposite or rarely whorled, the blades entire, the venation (including secondary nerves) sometimes very reduced and hardly discernible. Domatia variously formed but usually prominent on lower leaf surfaces in nerve axils. Staminate inflorescences few- to many-flowered cymes, axillary, pedunculate, often 2-armed. Staminate flowers regular, 4- or 5-merous; calyx limb lobed, toothed, or truncate; corolla infundibular with subimbricate lobes; stamens inserted on corolla tube, mostly included, longitudinally dehiscent; pistillodium included. Pistillate inflorescences 1- to many-flowered cymes, axillary, pedunculate, often 2-armed. Pistillate flowers regular, 4- to 10-merous; calyx limb lobed, toothed, or truncate, persistent in fruit; corolla infundibular to hypocrateriform with subimbricate lobes; staminodia of pistillate flowers inserted on corolla tube, short, included, usually as many as corolla lobes; style exserted, variously and often unequally branched, the number of branches usually equalling the number of corolla lobes; ovary inferior, 6- to many-loculed, the locules arranged in radiating double-files from center or apex of ovary; ovules solitary and pendulous in each locule. Fruits drupaceous with thin mesocarp, the 1-seeded pyrenes separate, vertical or variously radiating from apex or center of fruit.

Timonius is a genus of perhaps 180 species distributed from the Mascarene and Seychelle Islands to Ceylon and eastward throughout Malesia to tropical Australia, Micronesia, and the Tuamotu Islands. The type species (*Timonius timon*) and its closest relatives have been segregated as subg. *Timonius* (Darwin 1993). Twenty-nine other species are here assigned to subg. *Abbottia*.

Timonius subgenus **Abbottia** (F. v. Mueller) S. Darwin, comb. et stat. nov. *Abbottia* F. v. Mueller, Fragm. 9: 181. 1875, non *Abbotia* Rafinesque (Juncaginaceae), 1836.—TYPE: *Abbottia singularis* F. v. Mueller [≡*Timonius singularis* (F. v. Mueller) L. S. Smith].

Dioecious woody epiphytes or eventually terrestrial shrubs to tall forest trees. Stipules separate, imbricate, glabrous or pubescent, caducous or subpersistent. Leaves opposite, petiolate, the blades entire, chartaceous to thick-coriaceous; secondary nerves rarely well developed, usually submerged and indistinguishable from tertiary venation; veinlets forming groups of subparallel (reticuli-paxillate) areoles. Domatia usually present on lower leaf surfaces, cryptlike or pitlike, densely pubescent to glabrous; domatioid nectaries absent. Inflorescences dimorphic; staminate inflorescences 3- to 25-flowered; pistillate inflorescences 1- or sometimes 3- to 5-flowered. Inflorescence bracts and bracteoles reduced or absent. Calyx limb truncate to undulate. Corolla narrowly infundibular, glabrous within. Fruits obovoid to subglobose or subcylindrical, glabrous to sparsely pubescent, reddish to black when mature, the mesocarp thin. Pyrenes 15–350 (–500) per fruit, arranged in (4–) 6–8 (–12) double-files, vertical and subparallel or radiating from distal end of fruit.

Timonius subgenus *Abbottia* is reasonably well characterized by the reticuli-paxillate leaf venation with usually reduced secondary (lateral) nerves, usually solitary pistillate flowers, truncate calyx limbs, imbricate stipules, and, as far as known, an epiphytic habit; however, none of these characters is unique to the subgenus. At least one other species of *Timonius*, *T. epiphyticus* Elmer (Philippines) is a woody epiphyte, many other species have stipules that are imbricate in bud (likely the ancestral state for the genus), and 1-flowered pistillate inflorescences appear to have evolved independently in several species-groups of *Timonius* (e.g., subg. *Timonius*). Reduction of secondary nerves characterizes subg.

Abbottia, but some species included here on the evidence from other characters have rather prominent lateral nerves and relatively thin-textured leaf blades (e.g., *T. bougainvillensis*).

KEY TO THE SPECIES OF TIMONIUS SUBGENUS ABBOTTIA [Based on Reproductive and Vegetative Structures]

1. Corolla lobes with 3 conspicuous, irregularly elongate appendages arising from interior near apex; stipules deciduous subbasally and leaving a persistent and usually obvious collar at each node; Philippine Islands.
   2. Branchlets 2–3 mm wide toward apex; petioles 1–2 mm wide at middle; leaf blades stiff-chartaceous, 1.5–3 cm wide; staminate inflorescences 9- to 19-flowered, the peduncles ca. 1 mm wide; pistillate inflorescences (3-) 5- to 9-flowered. 3. *T. appendiculatus*.
   2. Branchlets 4–6 mm wide toward apex; petioles 3–4 mm wide at middle; leaf blades coriaceous, 3–5 cm wide; staminate inflorescences (5-) 7- to 9-flowered, the peduncles 2–3 mm wide; pistillate inflorescences 2- to 3-flowered. 19. *T. pachyphyllus*.
1. Corolla lobes lacking appendages, the interior distal surfaces plane to rugulose or longitudinally ribbed; stipules deciduous from very base, not leaving a nodal collar; eastern Borneo, Papuasia, Australia, Fiji, and Samoa.
   3. Base of leaf blades rounded. 18. *T. oktediensis*.
   3. Base of leaf blades acute or obtuse or truncate (sometimes narrowed but abruptly rounded in *T. trichanthus* or *T. belensis*).
      4. Pyrenes of fruit unequal in length, the lateral ones usually conspicuously shorter than the central, pendulous ones; fruits obovoid, widest above middle.
         5. Stipules adaxially densely pale-sericeous; leaf blades abaxially glabrous. 21. *T. pubistipulus*.
         5. Stipules adaxially glabrous (except for scattered secretory colleters) or pubescent with scattered hairs; leaf blades abaxially appressed-puberulent or sometimes glabrous.
            6. Pyrenes 15–25 in cross section of fruit; corolla of pistillate flowers 4-lobed; calyx tube of staminate flowers 1–1.5 mm long; corolla limb of staminate flowers 7–9 mm wide; stipules to 35 mm long; western New Guinea. 27. *T. trichocladus*.
            6. Pyrenes 30–40 in cross section of fruit; corolla of pistillate flowers 6-lobed; calyx tube of staminate flowers 2–4 mm long; corolla limb of staminate flowers 4–6 mm wide; stipules to 70 mm long; eastern New Guinea. 2. *T. amungwiwanensis*.
      4. Pyrenes of fruit roughly equal in length, radiating from the distal end of fruit or vertical and parallel; fruit globose to depressed-globose or cylindrical, broadest at middle (sometimes obovoid in *T. affinis*).
         7. Pyrenes vertical and parallel, the mature fruits usually strongly sulcate.
            8. Pistillate inflorescences 3- to 5-flowered cymes, the flowers rarely solitary.
               9. Branchlets glabrous; stipules to 130 mm long, to 15 mm wide; leaf blades thick-coriaceous, abaxially glabrous; domatia small, pitlike, scattered over abaxial surface of lamina; staminate inflorescences 5- to 7-flowered, the peduncles 6–10 mm long, ca. 3 mm wide, glabrous; calyx of staminate flowers to 4.5 mm long; calyx of pistillate flowers glabrous; fruits 4–5 mm long, 6–7 mm in diameter; extreme southeastern Papua. 29. *T. zuckianus*.
               9. Branchlets puberulent but soon glabrous; stipules to 60 mm long, to 8 mm wide; leaf blades chartaceous, abaxially glabrous to scabridulous; domatia appearing as tufts of hairs along costa or absent; staminate inflorescences 16- to 25-flowered, the peduncles 10–14 mm long, 1–1.5 mm wide, puberulent; calyx of staminate flowers to 2 mm high; calyx of pistillate flowers puberulent; fruits 3–3.5 mm long, 5–6.5 mm in diameter; north-central New Guinea. 13. *T. longifolius*.
            8. Pistillate inflorescences 1-flowered.
               10. Abaxial surface of stipules densely white- to yellow-strigillose to tomentulose or shaggy-hirsute; branchlets and inflorescences often (but not always) densely pubescent.

11. Branchlets densely pubescent with reddish hairs; stipules to 17 mm long, densely white- to yellow-strigillose; abaxial surfaces of leaf blades subvelutinous with spreading, reddish hairs; domatia absent; corolla tube of pistillate flowers 10–11 mm long; Papua. 9. *T. cryptophlebus*.
11. Branchlets glabrous to appressed- or spreading-pubescent with pale to brownish hairs; stipules usually more than 20 mm long, tomentose, hirsute, or strigillose (*T. dolichophyllus*); abaxial surfaces of leaf blades appressed- to spreading-pubescent, tomentulose, or strigillose; domatia distributed along costa or absent; corolla tube of pistillate flowers to 8 mm long.
12. Petioles hirsute; domatia absent; calyx of staminate flowers distally glabrous, pubescent toward base, the tube 0.5–1 mm long; peduncles of pistillate inflorescences 1–5 mm long; central New Guinea. 25. *T. subavenis*.
12. Petioles puberulent; domatia distributed along costa or scattered over abaxial surface of lamina; calyx of staminate flowers distally pubescent, the tube 2–3 mm long; peduncles of pistillate inflorescences (2–) 9–20 mm long.
13. Stipules shaggy-tomentulose; domatia pubescent, distributed along abaxial costa; eastern New Guinea (and Geelvink Bay region?). 26. *T. trichanthus*.
13. Stipules densely white- to yellow-strigillose; domatia glabrous, scattered over abaxial surface of leaf blade; central New Guinea. 10. *T. dolichophyllus*.
10. Abaxial surface of stipules glabrous to minutely appressed-puberulent to canescent; branchlets and inflorescences glabrous to scattered-puberulent.
14. Leaf blades rhombic-ovate, rounded to retuse at apex; corolla tube of staminate flowers ca. 0.5 mm wide; peduncles of pistillate inflorescences 2–4 mm long; hypanthium obturbinate, to 0.8 mm long and ca. 1 mm in diameter at anthesis; pyrenes to 20 in cross section of fruit; west-central New Guinea. 28. *T. wollastonii*.
14. Leaf blades elliptic to oblanceolate to obovate, sharply acute to acuminate (rarely obtuse) at apex; corolla tube of staminate flower ca. 1–2 mm wide (unknown in *T. heptamerus*); peduncles of pistillate inflorescences usually more than 4 mm long [(1–) 3–15 mm long]; hypanthium 1.5–3 mm long and (1.3–) 3–4 mm in diameter at anthesis; pyrenes ca. 30–50 in cross section of fruit.
15. Branchlets 3–4 mm wide toward apex, glabrous and often glaucous; staminate inflorescences 10- to 17-flowered, the corolla lobes 3–4 mm long; western New Guinea. 12. *T. lamii*.
15. Branchlets ca. 2 mm wide toward apex, glabrous to finely puberulent; staminate inflorescences (where known) 3- to 7- (13-)flowered, the corolla lobes to 3 mm long.
16. Leaf blades chartaceous to subcoriaceous; petioles slender, ca. 1 mm wide at middle; peduncles of staminate inflorescences (where known) mostly longer than wide; calyx tube of staminate flowers to 2 mm long; calyx of pistillate flowers truncate to irregularly denticulate.
17. Petioles usually not more than 10 mm long and 1 mm wide; peduncles of pistillate inflorescences glabrous; calyx of pistillate flowers glabrous or rarely puberulent, the corolla glabrous to appressed-puberulent outside, the hypanthium ca. 2 mm in diameter at anthesis, glabrous; New Guinea to Borneo. 4. *T. avenis*.
17. Petioles 10–15 mm long, 1–1.5 mm wide; peduncles of pistillate inflorescences puberulent; calyx of pistillate flowers glabrous to scattered-puberulent, the corolla densely pale-

strigillose outside, the hypanthium 3–3.5 mm in diameter at anthesis, subglabrous; western New Guinea (vicinity of Mt. Carstensz). 11. *T. heptamerus*.

16. Leaf blades thick-coriaceous; petioles stout, ca. 2 mm wide at middle; peduncles of staminate inflorescences wider than long; calyx tube of staminate flowers 2–3 mm long; calyx of pistillate flowers undulate to 7- or 8-lobed, the lobes ca. 0.5 mm long; western New Guinea (vicinity of Mt. Carstensz). 8. *T. carstensensis*.

7. Pyrenes oblique (sometimes subvertical in *T. belensis*), the mature fruits proximally tuberculate when dry due to shrinkage of pericarp against pyrenes (or fruits smooth in *T. secundiflorus*).

18. Secondary nerves prominulous on abaxial leaf surfaces, readily discernible, 11–13 (–15) on each side of costa; leaf blades oblong (occasionally elliptic to lanceolate); Solomon Islands (Bougainville, Kolombangara, Guadalcanal). 7. *T. bougainvillensis*.

18. Secondary nerves 5–8 on each side of costa or essentially indiscernible; leaf blades lanceolate to elliptic to obovate to suborbicular.

19. Fruits smooth, subglobose, not at all costate or tuberculate when dry; inflorescences usually glaucous when young; peduncles of staminate inflorescences slender, less than 1 mm wide; northern New Guinea. 23. *T. secundiflorus*.

19. Fruits conspicuously tuberculate-costate (at least toward base) when dry, glabrous but not glaucous; peduncles of staminate inflorescences usually more than 1 mm wide (ca. 0.5 mm wide in *T. pseudaffinis*).

20. Corolla tube of staminate flowers mostly 18–22 mm long, the lobes 7–15 mm long, ca. 5 mm wide, the limb to 30 mm in diameter; corolla lobes of pistillate flowers more than 5 mm long; calyx tube of pistillate flowers 3–6 mm long; fruits mostly larger than 15 mm long and 13 mm in diameter; Solomon Islands and New Britain. 14. *T. longitubus*.

20. Corolla tube of staminate flowers to 18 mm long, the lobes to 6 mm long and 3 mm wide, the limb to 12 mm in diameter (unknown in *T. neobritannicus*); corolla lobes of pistillate flowers to 5 mm long; calyx tube of pistillate flowers to 2 mm long; fruits to 13 mm long and 11 mm in diameter.

21. Stipules to 120 mm long and 20 mm wide; staminate inflorescences to 5 cm long, the peduncles 7–20 mm long; calyx tube of staminate flowers to 6 mm long; anthers 10 mm long; corolla tube of pistillate flowers mostly over 10 mm long; staminodia 4–5 mm long; fruits mostly 10–13 mm long; Santa Cruz Islands, Vanuatu, Fiji, Samoa. 1. *T. affinis*.

21. Stipules to 70 mm long and 15 mm wide; staminate inflorescences to 2 cm long, the peduncles usually less than 10 mm long; calyx tube of staminate flowers to 3 mm long; anthers to 6 mm long (unknown in *T. belensis* and *T. bismarckensis*); corolla tube of pistillate flowers to 8 mm long; staminodia to 3 mm long; fruits to 9 mm long.

22. Young branchlets, peduncles, and corollas of staminate and pistillate flowers densely spreading-puberulent; leaf blades pubescent abaxially with spreading to subappressed hairs scattered over lamina; stipules pale canescent abaxially and with longer, spreading, brownish hairs toward costa and base; northeastern New Guinea. 22. *T. scabriflorus*.

22. Young branchlets, peduncles, corollas of staminate and pistillate flowers, and abaxial surfaces of leaf blades glabrous to sparsely and minutely appressed-puberulent; stipules glabrous to sparsely appressed-puberulent.

23. Corollas short-tubular to infundibular, the corolla tube of staminate flowers 6–7 mm long and 2 mm wide; corolla tube

of pistillate flowers 2–6 mm long and 2.5–3 mm wide, the hypanthium 3.5–5 mm long and 4–5 mm in diameter at anthesis; New Guinea. 5. *T. belensis*.

23. Corollas narrowly infundibular to hypocrateriform, the corolla tubes of staminate flowers usually more than 8 mm long and less than 2 mm wide; corolla tubes of pistillate flowers usually 1.5–3 mm wide, the hypanthium rarely (*T. bismarckensis* and *T. singularis*) more than 3 mm long and 3.5 mm in diameter.
    24. Stipules 10–25 mm long; petioles slender, usually less than 1 mm wide at middle; peduncles less than 1 mm wide; Solomon Islands. 20. *T. pseudaffinis*.
    24. Stipules 15–70 mm long; petioles 1–3 mm wide; peduncles 1.5–2 mm wide.
        25. Domatia pocketlike, solitary or paired near abaxial base of leaf blade or sometimes distributed along costa or rarely scattered over abaxial surface of leaf blades; leaf blades to 3 cm wide; corolla tube of pistillate flowers ca. 4 mm long and 1.5 mm wide, the lobes 6; fruits ca. 4 mm long and 6 mm in diameter when dry; New Britain and Papuan Islands. 17. *T. neobritannicus*.
        25. Domatia absent or pocketlike and distributed along costa, or small and pitlike distributed over abaxial surface of leaf blades; leaf blades often more than 4 cm wide; corolla tube of pistillate flowers often more than 4 mm long and 1.5 mm wide, the lobes 6–11; fruits 4–8 mm long and 6–10 mm in diameter.
            26. Leaf blades elliptic, abaxially glabrous to sparsely appressed-puberulent; peduncles of pistillate inflorescences 2–5 (–8) mm long; corolla tube of pistillate flowers 6–8 mm long; southeastern Papua. 15. *T. meridionalis*.
            26. Leaf blades elliptic to obovate, abaxially glabrous; peduncles of pistillate inflorescences usually more than 5 (often 10–20) mm long; corolla tube of pistillate flowers to 6 mm long.
                27. Petioles 10–30 mm long; peduncles of pistillate inflorescences ca. 5 mm long at anthesis (to 10 mm in fruit); corolla tube of pistillate flowers 4–5 mm long and 2–3 mm wide, the hypanthium ca. 2 mm long at anthesis; Queensland. 24. *T. singularis*.
                27. Petioles 5–15 mm long; peduncles of pistillate inflorescences 5–15 mm long at anthesis (to 20 mm in fruit); corolla tube of pistillate flowers usually more than 5 mm long, 1–2 mm wide, the hypanthium usually 3–4 mm long at anthesis.
                    28. Stipules to 45 mm long; domatia usually absent, or small and pitlike and distributed over abaxial surface of leaf blades; calyx tube of pistillate flowers ca. 0.5 mm long, the corolla limb 5–6 mm in diameter; Bismarck Archipelago. 6. *T. bismarckensis*.

28. Stipules to 70 mm long; domatia small open pits distributed over abaxial surface of leaf blades, and often also pocket- or porelike along abaxial leaf margins; calyx tube of pistillate flowers 0.5–1.5 mm long, the corolla limb 6–10 mm in diameter; south-central and western New Guinea. 16. *T. modestus*.

Key to the Species of Timonius Subgenus Abbottia [Based on Vegetative Structures and Geographical Distribution]

1. Leaf blades lanceolate- to elliptic-oblong, the lateral nerves prominulous on abaxial surface; Solomon Islands (Bougainville, Kolombangara, Guadalcanal). 7. *T. bougainvillensis*.
1. Leaf blades obovate to elliptic, the lateral nerves not raised on abaxial surface, or if prominulous then the blades not at all oblong.
    2. Leaf blades tapering to a rounded or truncate base.
        3. Branchlets glabrous except sometimes pubescent above stipule scars.
            4. Leaf domatia (in axils of lateral nerves on abaxial surface) conspicuously pubescent; northern New Guinea (Sepik River region). 13. *T. longifolius*.
            4. Leaf domatia glabrous or subglabrous; central New Guinea (Ok Tedi headwaters). 18. *T. oktediensis*.
        3. Branchlets spreading-puberulent with brownish hairs.
            5. Stipules minutely puberulent adaxially; southeastern New Guinea. 9. *T. cryptophlebus*.
            5. Stipules glabrous adaxially.
                6. Stipules gray- and appressed-pubescent abaxially; northeastern New Guinea. 22. *T. scabriflorus*.
                6. Stipules brown- or stramineous- and spreading-pubescent abaxially; New Guinea. 26. *T. trichanthus*.
    2. Leaf blades obtuse to acute or attenuate at base.
        7. Stipules pubescent adaxially, at least toward center or base.
            8. Leaf blades velutinous-pubescent abaxially; southeastern New Guinea. 9. *T. cryptophlebus*.
            8. Leaf blades glabrous or rarely with a few scattered hairs abaxially.
                9. Branchlets (terminal internode) 4–6 mm wide; Solomon Islands (Bougainville to Guadalcanal) and New Britain. 14. *T. longitubus*.
                9. Branchlets (terminal internode) 2–4 mm wide.
                    10. Stipules adaxially scattered- and appressed-puberulent; New Britain, New Ireland, and Manus Island. 6. *T. bismarckensis*.
                    10. Stipules adaxially sericeous to densely strigillose.
                        11. Domatia open and pitlike and distributed over abaxial surface of leaf blades; New Guinea (Snow Mountains to southeastern Papua). 21. *T. pubistipulus*.
                        11. Domatia partially covered (cryptlike or pocketlike), distributed over abaxial surface of leaf blades and often also along sides of costa, sometimes absent; Philippine Islands (Luzon, Mindoro, Biliran, Mindanao). 3. *T. appendiculatus*.
        7. Stipules adaxially glabrous except for scattered secretory colleters (although the branchlets frequently pubescent in stipule axils).
            12. Leaf blades rhombic-ovate, rounded to retuse at apex; western central New Guinea. 28. *T. wollastonii*.
            12. Leaf blades elliptic to oblanceolate or obovate, obtuse to acuminate or mucronate at apex, or if rounded (*T. pachyphyllus*) then not rhombic in outline.
                13. Terminal buds (abaxial surface of stipules) conspicuously and often densely pubescent.

14. Terminal buds densely white-, gray-, or yellow-sericeous, the stipule surface hidden by indument.
    15. Leaf blades 3 to 4 times as long as wide, narrowly elliptic to oblanceolate; terminal buds yellowish pubescent; central New Guinea (Star Mountains to Fly River). 10. *T. dolichophyllus*.
    15. Leaf blades 2 to 2.5 times as long as wide, broadly elliptic to obovate; terminal buds white- or gray-pubescent; northeastern New Guinea. 22. *T. scabriflorus*.
14. Terminal buds brown- to stramineous-pubescent with appressed or spreading hairs, the stipule surface visible beneath indument.
    16. Petioles 18–40 mm long; leaf blades broadly elliptic to obovate; stipules 40–100 mm long; Solomon Islands (Bougainville to Guadalcanal) and New Britain. 14. *T. longitubus*.
    16. Petioles 4–18 mm long; leaf blades elliptic to oblanceolate; stipules 10–45 mm long (in *T. longifolius* to 60 mm long).
        17. Branchlets glabrescent, or at least much less densely pubescent than the terminal buds; New Guinea (Sepik River region). 13. *T. longifolius*.
        17. Branchlets spreading-pubescent, usually about as densely pubescent as the terminal buds.
            18. Domatia absent; north-central New Guinea. 25. *T. subavenis*.
            18. Domatia present, appearing as tufts of hairs along costa.
                19. Terminal buds densely spreading-pubescent; leaf blades abaxially pubescent with appressed hairs, or often velutinous with spreading indument; New Guinea. 26. *T. trichanthus*.
                19. Terminal buds sparsely appressed-pubescent; leaf blades abaxially appressed-pubescent; western New Guinea (Utakwa River to Mt. Carstensz). 11. *T. heptamerus*.

13. Terminal buds (abaxial surface of stipules) glabrous or subciliate to minutely appressed-puberulent with scattered hairs.
    20. Domatia pocket- or slitlike along sides of costa, sometimes also pitlike and distributed over abaxial surface of leaf blades, or cryptlike along costa and on abaxial leaf surface (*T. amungwiwanensis*).
        21. Branchlets (terminal internodes) spreading-puberulent, often minutely so.
            22. Plants of southeastern New Guinea and the Papuan Islands. 15. *T. meridionalis*.
            22. Plants of northern and western New Guinea.
                23. Leaf blades obovate to elliptic, 4–13 cm long, 2–6 cm wide, abaxially glabrous; New Guinea. 16. *T. modestus*.
                23. Leaf blades elliptic to oblanceolate or sometimes obovate, 1–11 (–15) cm long, 0.5–3 (–4.5) cm wide, abaxially glabrous or pubescent on costa.
                    24. Branchlets (terminal internodes) 3–4 mm wide at middle; western New Guinea (Wissel Lakes and Lake Habbema). 27. *T. trichocladus*.
                    24. Branchlets (terminal internodes) 1–2 (–3) mm wide at middle.
                        25. Plants of north-central and northeastern New Guinea. 23. *T. secundiflorus*.
                        25. Plants of far western New Guinea (Vogelkop Peninsula and Geelvink Bay region). 4. *T. avenis*.
        21. Branchlets (terminal internodes) glabrous except often ciliate above stipule scars.
            26. Leaf blades pubescent on abaxial costa.

27. Leaf blades 2.5–7 cm wide; Santa Cruz Islands, Vanuatu, Fiji, Samoa. 1. *T. affinis*.
27. Leaf blades 0.5–4.5 cm wide.
28. Lateral nerves and veinlets prominulous on abaxial leaf surfaces (seen in dried specimens); southeastern New Guinea. 2. *T. amungwiwanensis*.
28. Lateral nerves and veinlets submerged, the abaxial leaf surfaces plane.
29. Leaf blades chartaceous to thin-coriaceous; eastern Borneo to eastern New Guinea. 4. *T. avenis*.
29. Leaf blades thick- and stiff-coriaceous; western New Guinea (vicinity of Mt. Carstensz). 8. *T. carstensensis*.
26. Leaf blades glabrous on abaxial costa, although sometimes puberulent over domatia.
30. Lateral nerves and veinlets prominulous on abaxial leaf surfaces (seen in dried specimens); leaf blades either chartaceous or oblanceolate.
31. Leaf blades 2.5–7 cm wide, elliptic to obovate; Santa Cruz Islands, Vanuatu, Fiji, Samoa. 1. *T. affinis*.
31. Leaf blades 1–3.5 cm wide, oblanceolate; Solomon Islands (Rob Roy to San Cristobal). 20. *T. pseudaffinis*.
30. Lateral nerves and veinlets submerged, the abaxial leaf surfaces plane, or if venation discernible then the blades coriaceous or elliptic to obovate.
32. Abaxial surfaces of leaf blades drying distinctly paler than adaxial surfaces, or if leaves concolorous then the abaxial venation indiscernible.
33. Leaf blades thick- and stiff-coriaceous; western New Guinea (vicinity of Mt. Carstensz). 8. *T. carstensensis*.
33. Leaf blades chartaceous to coriaceous.
34. Plants of Australia (Cape York Peninsula). 24. *T. singularis*.
34. Plants of Papuasia (Borneo to southeastern New Guinea).
35. Plants of southeastern New Guinea and the Papuan Islands. 15. *T. meridionalis*.
35. Plants of eastern Borneo, east-central New Guinea, and intervening land areas.
36. Plants of north-central and northeastern New Guinea (Sepik River region). 23. *T. secundiflorus*.
36. Plants of eastern Borneo to eastern New Guinea. 4. *T. avenis*.
32. Abaxial and adaxial leaf surfaces drying about the same shade, or if paler abaxially then the venation (at least lateral nerves) discernible.
37. Terminal internodes (and sometimes also petioles and stipules) glaucous; western New Guinea. 12. *T. lamii*.
37. Terminal internodes not glaucous.
38. Terminal internodes (3–) 4–5 mm wide at middle; Lake Habbema (western New Guinea) and highlands of eastern New Guinea. 5. *T. belensis*.
38. Terminal internodes 1–3 mm wide at middle.
39. Leaf blades 1–3 cm wide, oblanceolate to narrowly elliptic; New Britain, Umboi Island, and Rossel Island. 17. *T. neobritannicus*.
39. Leaf blades (1–) 2–9 cm wide, obovate to elliptic.

40. Leaf blades chartaceous; stipules 15–40 mm long; southeastern New Guinea and Papuan Islands. 15. *T. meridionalis.*
40. Leaf blades coriaceous; stipules 25–70 mm long; New Guinea. 16. *T. modestus.*

20. Domatia absent, or exclusively pitlike and distributed over abaxial surface of leaf blades, not associated with costa.
41. Terminal internodes (and sometimes also petioles and stipules) glaucous; western New Guinea. 12. *T. lamii.*
41. Terminal internodes glabrous or puberulent, but not glaucous.
42. Terminal internodes 4–6 mm wide at middle.
43. Branchlets puberulent; western New Guinea (Wissel Lakes and Lake Habbema). 27. *T. trichocladus.*
43. Branchlets glabrous (although sometimes puberulent above stipule scars).
44. Leaf blades narrowed toward the apex but rounded at the tip; Philippine Islands (Luzon). 19. *T. pachyphyllus.*
44. Leaf blades acute to sharply acuminate or mucronate at apex.
45. Stipules 60–130 mm long; petioles 12–55 mm long; leaf blades acuminate-mucronate at apex; islands of southeastern Papua New Guinea. 29. *T. zuckianus.*
45. Stipules 15–50 mm long; petioles 5–25 mm long; leaf blades acute to acuminate at apex; Lake Habbema (western New Guinea) and highlands of eastern New Guinea. 5. *T. belensis.*
42. Terminal internodes 1–3 mm wide at middle.
46. Leaf blades acuminate-mucronate at apex, the mucro 2–5 mm long; islands of southeastern Papua New Guinea. 29. *T. zuckianus.*
46. Leaf blades sharply to bluntly acute or acuminate at apex, but not mucronate.
47. Branchlets pubescent with hairs 0.3–0.5 mm long; western New Guinea (Wissel Lakes and Lake Habbema). 27. *T. trichocladus.*
47. Branchlets glabrous, or minutely puberulent with hairs ca. 0.1 mm long, or pubescent with longer hairs only above stipule scars.
48. Abaxial surface of leaf blades plane, the minor venation immersed and not discernible.
49. Plants of Australia (Cape York Peninsula). 24. *T. singularis.*
49. Plants of eastern Borneo, Papuan Islands, and intervening land areas.
50. Plants of southeastern New Guinea and the Papuan Islands. 15. *T. meridionalis.*
50. Plants of eastern Borneo, eastern New Guinea, and intervening land areas. 4. *T. avenis.*
48. Abaxial surface of leaf blades striate, the minor venation prominulous and discernible.
51. Leaf blades to 3.5 cm wide, narrowly oblanceolate, the fine venation prominulous on abaxial leaf surfaces; Solomon Islands (Rob Roy to San Cristobal). 20. *T. pseudaffinis.*
51. Leaf blades usually more than 3.5 cm wide, elliptic to obovate, if oblanceolate then the venation hardly discernible on abaxial surfaces.
52. Leaf blades chartaceous; stipules 50–120 mm long, rarely shorter when mature; Santa Cruz Islands, Vanuatu, Fiji, Samoa. 1. *T. affinis.*

52. Leaf blades coriaceous, or if chartaceous then stipules 15–50 mm long.
    53. Plants of southeastern New Guinea and the Papuan Islands. 15. *T. meridionalis*.
    53. Plants of eastern Borneo to Bismarck Archipelago and intervening land areas.
        54. Plants of the Bismarck Archipelago (New Britain, New Ireland, and Manus Island). 6. *T. bismarckensis*.
        54. Plants of New Guinea westward to Borneo.
            55. Leaf blades 2–6 cm wide, obovate to elliptic; New Guinea. 16. *T. modestus*.
            55. Leaf blades 0.5–3 (–4.5) cm wide, narrowly elliptic to oblanceolate; eastern Borneo to eastern New Guinea. 4. *T. avenis*.

1. Timonius affinis A. Gray, Proc. Amer. Acad. Arts 4: 36. 1858.—TYPE: FIJI. Without locality, 1840, *U.S. Expl. Exped. s.n.* ♀ (holotype: US-62264!; isotype: GH!).

Epiphytic shrubs to trees to 13 m tall. In var. *affinis*, bark pale brown and scaly, the wood hard; nature of bark and wood unknown in vars. *samoensis* and *sapotifolius*. Branchlets 2–3 mm wide toward apex, gray to brownish, glabrous. Stipules long-lanceolate to linear, to 120 mm long and 20 mm wide, acuminate, glabrous abaxially, glabrous adaxially except for colleters near base. Petioles (10–) 15–45 mm long, 1–2 mm wide at middle, semi- to subterete, glabrous; leaf blades 5–14 cm long, 2–7 cm wide, elliptic to obovate, acute to sharply acuminate at apex, attenuate at base, chartaceous to thin-coriaceous, glabrous above, glabrous or with scattered, appressed hairs over lamina and on nerves beneath, the costa prominulous to canaliculate above, elevated beneath, the secondary nerves usually discernible, 5–8 on each side of costa, spreading, the veinlets prominulous or submerged in lamina, paxillate; domatia absent or present and pocketlike in axils of secondary nerves, glabrous. Staminate inflorescences (1-) 3- to 7-flowered, at anthesis to 5 cm long and 6 cm wide; peduncles 7–20 mm long, about 2 mm wide, compressed, glabrous; bracts absent. Staminate flowers: calyx limb cupular, truncate to undulate, glabrous, the tube to 6 mm long from base; corolla narrowly infundibular, the tube (7–) 12–18 mm long, ca. 2 mm wide, glabrous to sparsely scattered-puberulent outside, glabrous within, the lobes 4, to 6 mm long and 3 mm wide, lanceolate, acute, papillose, the corolla limb to 12 mm in diameter; stamens 4, the anthers 10 mm long, 0.5–1 mm wide, linear-sagittate, glabrous; pistillodium ca. 8 mm long, filiform, undivided, glabrous. Pistillate inflorescences 1- (2-)flowered; peduncles 8–25 (in fruit to 40) mm long, 1–2 mm wide, compressed, glabrous. Pistillate flower: calyx limb cupular, truncate, glabrous, the tube 0.5–2 mm long; corolla narrowly infundibular, the tube (7–) 10–15 mm long, about 2 mm wide, glabrous outside, glabrous within, the lobes 5–8, 4–5 mm long, 1–2 mm wide, lanceolate, acute, papillose, the corolla limb 8–12 mm in diameter; staminodia 5–6 (–8), 4–5 mm long, 0.5 mm wide, linear-sagittate, glabrous; hypanthium 3–5 mm long, 2.5–5 mm in diameter, depressed-globose, glabrous, the disc glabrous, often elevated beyond calyx in fruit, the style to 18 mm long, about 1 mm wide near middle, irregularly branched, glabrous. Fruits (8–) 10–13 mm long, 8–11 mm in diameter, obovoid to subglobose, tuberculate to ribbed, glabrous, purple to black when mature; pyrenes (18–) 50–350, subequal, obliquely oriented. Fig. 7e, f.

Timonius affinis is the easternmost element of subg. *Abbottia*, extending from the Santa Cruz Islands and Vanuatu to Fiji and Samoa. It can be recognized by the large and often obovoid fruits with often very numerous pyrenes, and the large, glabrous stipules. *Timonius pseudaffinis*, from the Solomon Islands, has passed for *T. affinis* in herbaria, but has smaller fruits (5–9 mm in diameter) and stipules (the largest seen only 25 mm long), and more slender branchlets, petioles, and peduncles as well as consistently oblanceolate leaf blades 1–3.5 cm wide. Three varieties of *T. affinis* are recognized here, one of them in the Santa Cruz Islands, Vanuatu, and Fiji, one endemic to Fiji, and a third endemic to Samoa.

Key to the Varieties of Timonius affinis

1. Corollas of pistillate flowers (5-) 6-lobed; calyx tube (staminate and pistillate flowers) 2–4 mm long; lateral nerves usually distinguishable from veinlets.
   2. Veinlets (or at least lateral nerves) prominulous on lower leaf surfaces; fruits with (15–) 20–30 pyrenes in longitudinal section; Santa Cruz Islands, Vanuatu, and Fiji. 1a. *T. affinis* var. *affinis*.
   2. Veinlets (including lateral nerves) submerged on lower leaf surfaces; fruits with 6–10 pyrenes in longitudinal section; Fiji. 1c. *T. affinis* var. *sapotifolius*.
1. Corollas of pistillate flowers 8-lobed; calyx tube ca. 1 mm long; lateral nerves of leaf blades hardly differentiated from veinlets; Samoa. 1b. *T. affinis* var. *samoensis*.

1a. Timonius affinis var. affinis.

Bark pale brown and scaly, the wood hard. Leaf venation usually prominulous beneath, the lateral nerves thicker and more prominent than tertiary nerves or veinlets. Calyx tube of staminate and pistillate flowers 2–4 mm long. Corolla of pistillate flowers (5-) 6-lobed. Fruits with (50–) 90–350 pyrenes.

Phenology. Flowering and fruiting throughout the year.

Distribution. Fiji, Santa Cruz Islands, and Vanuatu (one collection from Espiritu Santo); dense, dry, or thin and rocky forests, or thickets of crests and ridges; 0–1160 m.

Vernacular names. That the species is suggestive of mangroves (*ndongo* or *tiri*) is indicated by the following Fijian names recorded on herbarium labels: *tiri vanua, ndongo, ndongo ni vanua, ndongo ni lekutu, ndongo tangane, ndongo vuto*, and *ndongondongo*; other Fijian names recorded only once are *mokamoka* (Namosi) and *naka pinda* (Naitasiri). The name *simidi* (Kwara'ae) is reported from the Santa Cruz Islands.

Additional Specimens Examined. **Santa Cruz Islands**. Ndeni [Santa Cruz]: Carlisle Bay area, *BSIP 17019* ♀ (K, L, LAE, SING).—Vanikoro: W side Ambi River valley, Hailings Road, *BSIP 1787* ♀ (K, L, LAE, SING); S coast at Emwa, *BSIP 1696* ♀ (L, LAE); without locality, *Kajewski 623* ♀ (A, BRI). **Vanuatu** [New Hebrides]. Espiritu Santo: NE of Cotamtam, trail to Nokowoula, *Roy. Soc. Exped. (Veillon) 4538* ♀ (A, L). **Fiji**. Viti Levu. Mba: eastern slope of Mt. Koroyanitu, Mt. Evans

Range, *Smith 4159* ♂ (A, BISH, BRI, L, SUVA); upper slopes of Mt. Koromba, *Smith 4640* ♀ (A, BISH, BRI, L); Mt. Nanggaranambuluta (Lomalangi), E of Nandarivatu, *Gillespie 4292* ♀ (BISH), *Gillespie 4342* ♂ (BISH), *Greenwood 873* ♀ (A, K), *Greenwood 873A* ♂ (A, K), *Smith 4763* ♀ (A, BISH, BRI, L, SUVA); Nandarivatu, above swimming pool, *Fiji Dept. Agr. 8509* (BRI); Tavua, *Webster & Hildreth 14209* ♀ (BISH, DAV, MASS); Mt. Tomanivi (Mt. Victoria), *Fiji Dept. Agr. 12719 (Melville & Parham 7108)* ♀ (A, K, L, SUVA). Nandronga and Navosa: northern portion of Rairaimatuku Plateau, between Nandrau and Nanga, *Smith 5507* ♀ (A, BISH, BRI, L). Serua: Namboutini, *Fiji Dept. Agr. 13712* (SUVA). Namosi: hills N of Wainavindrau Creek between Korombasambasanga Range and Mt. Naitarandamu, *Smith 8409* ♀ (BISH, GH, L, SUVA); Mt. Naitarandamu, *Gillespie 3106* ♀ (BISH); slopes of Mt. Voma, *Gillespie 2890* ♀ (BISH); hills E of Wainikoroiluva River, near Namuamua, *Smith 9058* ♀ (BISH, GH, L, SUVA); vicinity of Namosi Village, *Gillespie 2620* ♀ (A, BISH). Naitasiri: Taunaisali, Wainimala River Valley, *St. John 18339* ♀ (A, BISH); vicinity of Nasinu, *Gillespie 3552* ♀ (BISH, BRI), *Gillespie 3755.5* ♂ (BISH). Tailevu: hills E of Wainimbuka River, vicinity of Ndakuivuna, *Smith 7070* ♀ & ♂ (BISH, GH, L, RSA, SUVA). Rewa: quarry near Lami, *Gillespie 4613* ♀ (BISH). Province unknown: *Parks 20891* ♀ (BISH, L, SUVA).—Kandavu: Mt. Mbuke Levu, *Fiji Dept. Agr. 2988* (SUVA).—Ovalau: hills W of Lovoni Valley, on ridge S of Mt. Korolevu, *Smith 7526* ♀ (BISH, GH, L, SUVA); vicinity of Levuka, *Gillespie 4512* ♂ (BISH).—Koro: eastern slope of main ridge, *Smith 955* ♂ (BISH, GH).—Ngau: hills E of Herald Bay, inland from Sawaieke, *Smith 7777* ♀ (BISH, GH, L, SUVA).—Vanua Levu. Mathuata: Mathuata Range, *Smith 6794* ♀ (A, BISH, BRI, L, SUVA). Mathuata-Thakaundrove boundary: crest of Korotini Range, between Navitho Pass and Mt. Ndelaikoro, *Smith 547* ♀ (BISH, GH). Thakaundrove: Mt. Mbatini, *Smith 691* ♀ (BISH, GH); Mt. Kasi, Yanawai River region, *Smith 1764* ♀ (BISH, GH); ridge above Ndwara, Yanawai River, *Fiji Dept. Agr. 15751* ♀ (BISH, SUVA).—Taveuni: trail from Somosomo, *Gillespie 4817* (BISH); borders of lake E of Somosomo, *Smith 862* ♂ (BISH, GH).

Timonius affinis var. *affinis* is variously reported as a woody epiphyte, climber, shrub, or banyonlike tree. The faintly fragrant flowers have corollas with the tube at first greenish yellow but becoming purple outside and with pale yellow or white lobes.

1b. Timonius affinis var. **samoensis** S. Darwin, var. nov.—Type: Samoa. Savaii: above Sili in medium wet forest at about 400 m, 11 Nov 1931, *Christophersen 3246* ♀ (holotype: BISH-178059!; isotypes: A! BISH-178061!).

Idem ac varietas affinis sed nervis secundariis abaxialibus comparate obscuris et in lamina immersis, petiolis comparate latis (subalatis), calycis tubo brevi, et lobis corollae femineae 8 non 6 differt.

Leaf venation submerged in lamina beneath, the lateral nerves not any thicker or more prominent than tertiary nerves or veinlets. Calyx tube of staminate and pistillate flowers ca. 1 mm long. Corolla of pistillate flowers 8-lobed. Fruits with 80–150 pyrenes.

Phenology. Collected in flower and fruit sporadically throughout the year.

Distribution. Samoa; montane forests; 400–650 m.

Vernacular Name. *Aloaloa tane* (Parham, 1972, p. 14).

Additional Specimens Examined. **Samoa**. Savaii: Tuafa, *Reinecke 388* ♀ (BISH).—Upolu: top of Mt. Mariota near Aleisa, *Whistler 763* ♂ & ♀ (A, BISH); rim of crater SE of Mt. Mariota, *Whistler 1915* ♀ (BISH), *Whistler 1916* ♂ (BISH); above Togitogiga, *Whistler 3928* ♂ (BISH); S slope Mt. LePue, *Cox 192* ♀ (GH).

All available collections of *T. affinis* from Samoa are assignable to this variety. The type collection, bearing immature fruits, was taken from an epiphyte on a tall tree; other collections are said to be taken from epiphytes, "stranglers," or trees to 5 m tall. The flowers (staminate and pistillate) are yellowish to white.

1c. Timonius affinis var. **sapotifolius** (A. Gray) Fosberg, Sargentia 1: 121. 1942 (*'sapotaefolius'*). *Timonius sapotaefolius* A. Gray, Proc. Amer. Acad. Arts 4: 35. 1858.—TYPE: FIJI. Without locality [Ovalau?], 1840, *U.S. Expl. Exped. s.n.* ♀ (holotype: US-62268!; isotypes: GH! K!).

Leaf venation plane or submerged on lower surfaces of blades, the lateral nerves thicker and more prominent than tertiary nerves or veinlets. Calyx tube of staminate and pistillate flowers 2–4 mm long. Corolla of pistillate flowers 6-lobed. Fruits with 18–30 (–60) pyrenes.

Phenology. Collected in fruit in October.

Distribution. Fiji, Ovalau; 90–350 m.

Vernacular Name. *Mbaka loa* (*Fiji Dept. Agr. 17018*).

ADDITIONAL SPECIMENS EXAMINED. **Fiji**. OVALAU: wooded hills overlooking Levuka, *Gillespie 4468* (BISH); Waililevu, *Fiji Dept. Agr. 17018* ♀ (BRI).

The type collection of *T. affinis* var. *sapotifolius* was obtained in Fiji, without further locality; the fact that the only two other collections known to represent this variety are from Ovalau suggests that the Exploring Expedition material may also have come from that island, which was the expedition's principal base in Fiji (Smith & Darwin 1988). Staminate individuals are not yet available for study; the collection *Fiji Dept. Agr. 17018* was made from an epiphyte on *Inocarpus fagifer* growing on an open hillside.

2. Timonius amungwiwanensis S. Darwin, sp. nov.—TYPE: PAPUA NEW GUINEA. Morobe: Mt. Amungwiwa, bordering alpine grasslands, 3050 m, 12 Jun 1976, *van Royen 11085* ♀ (holotype: BISH!; isotype: LAE!).

Timonius ex affinitate T. pubistipuli S. Darwin et T. trichocladi Merr. & Perry, sed a priore stipulis adaxialiter glabris et foliorum laminis in costa abaxiali pilis appressis dispersis rigidis pubescentibus distinctus; a posteriore pyrenis in fructu numerosis et stipulis majoribus et lobis corollae femineae 6 non 4 differt.

Understory trees to 6 m tall, perhaps epiphytic when young. Bark pale to dark brown; wood straw-colored. Branchlets 2–3 mm wide toward apex, grayish, glabrous. Stipules to 70 mm long, 10 mm wide, lanceolate, acuminate, glabrous abaxially, glabrous adaxially except for colleters near base. Petioles 5–20 mm long, ca. 2 mm wide at middle, semi- to subterete, glabrous; leaf blades 4–9 cm long, 2–4.5 cm wide, elliptic to somewhat obovate, sharply to bluntly acute to acuminate at apex, attenuate at base, thick-coriaceous, glabrous above, puberulent with scattered, stiff, appressed hairs on costa and often also toward apex beneath, the costa narrowly canaliculate above, elevated beneath, the secondary nerves hardly discernible, 5–8 on each side of costa, spreading, the veinlets prominulous, paxillate; domatia on lamina and along costa beneath, cryptlike with a small ostiole. Staminate inflorescences 3-flowered, at anthesis ca. 1.5 cm long, ca. 1 cm wide; peduncles 2–4 mm long, 1–1.5 mm wide, glabrous; bracts usually absent but occasionally discernible, ca. 0.5 mm long and broadly deltate, or rarely 2 mm long and subulate. Staminate flowers: calyx limb cupular, truncate to undulate, glabrous, the tube 2–4 mm long from base; corolla infundibular, white, the tube ca. 7 mm long, ca. 1.5 mm wide, glabrous outside, glabrous within, the lobes 4, 3–4 mm long, 1–1.5 mm wide, ovate-lanceolate, acute, papillose, the corolla limb 4–6 mm in diameter; stamens 4, the filaments short, the anthers ca. 4.5 mm long, ca. 0.5 mm wide, linear, glabrous; pistillodium ca. 5 mm long, filiform, glabrous. Pistillate inflorescences 1-flowered; peduncles 2–3 mm long, ca. 1 mm wide, glabrous; bracts indistinguishable. Pistillate flowers: calyx limb cupular, truncate, glabrous, the tube ca. 1 mm long; corolla infundibular, white, the tube ca. 4 mm long, ca. 1.5 mm wide, glabrous outside, glabrous within, the lobes 6, 3–4 mm long, 1–1.5 mm wide, lanceolate, acute, papillose, the corolla limb ca. 8 mm in diameter; staminodia 6, ca. 3 mm long, ca. 0.5 mm wide, linear, glabrous; hypanthium 4–5 mm long, 2–3 mm in diameter, ellipsoid, glabrous, the disc glabrous, the style 6-armed, 3–4 mm long, 0.5–1 mm wide near middle, glabrous. Fruits 6–8 mm long, 6–10 mm in diameter, obovoid, somewhat laterally compressed, glabrous, red when mature; pyrenes 30–40 in 6 radiating double-files, unequal, subvertical. Figs. 7b, 8.

Phenology. Collected in flower and fruit from June to October.

Distribution (Fig. 13). Highlands of southeastern Papua New Guinea; subalpine forests; 2200–3800 m.

ADDITIONAL SPECIMENS EXAMINED. **Papua New Guinea**. [*Northeastern New Guinea*.] MOROBE: Mt. Amungwiwa, *Corlett 25162* (LAE), *Corlett 25164* ♂ (LAE), *Moi 107* ♀ (A, K, L, LAE).—[*Papua*.] CENTRAL: Murray Pass, *Brass 4705* ♂ (A, BO, BRI), *NGF 36847* ♂ (A, CANB, E, K, L, LAE).

Timonius amungwiwanensis appears to be a rare species in *Dacrycarpus*-dominated and usually stunted forest. It is within the geographical range of *T. pubistipulus*, and resembles that species in having obovoid fruits with subvertically oriented pyrenes but differs in having adaxially glabrous stipules and leaf blades with stiff, appressed, scattered hairs on the abaxial costa. *Timonius amungwiwanensis* differs from *T. trichocladus* in having a greater number of pyrenes per fruit (30–40 versus ca. 24, respectively), six rather than four pistillate corolla lobes, and larger stipules, among other characters.

3. **Timonius appendiculatus** Merrill, Philipp. J. Sci. 4: 327. 1909.—TYPE: PHILIPPINES. Luzon: Rizal Prov., vicinity of Antipolo, Jan 1907, *Philipp. Bur. Sci. 2164* ♂ (lectotype, designated by Elmer, 1911: PNH, destroyed; isolectotypes: GH! PNH!).

Epiphytic shrubs or trees to 25 m tall and 20 cm dbh; nature of bark and wood unknown. Branchlets 2–3 mm wide toward apex, pale gray, at length quadrangular, glabrous but ciliate above stipule scars. Stipules to 30 mm long, 10 mm wide, imbricate, oblong-lanceolate, acute, glabrous abaxially, adaxially densely appressed-strigose toward center and base among colleters, glabrous toward margin, caducous but the very bases persistent as a low collar. Petioles 8–20 mm long, 1–2 mm wide at middle, semiterete but winged by base of leaf blade, glabrous; leaf blades 5–11 cm long, 1.5–3 cm wide, narrowly elliptic to subfalcate, bluntly acute at apex, long attenuate at base, stiff-chartaceous, glabrous above and beneath, the costa plane to

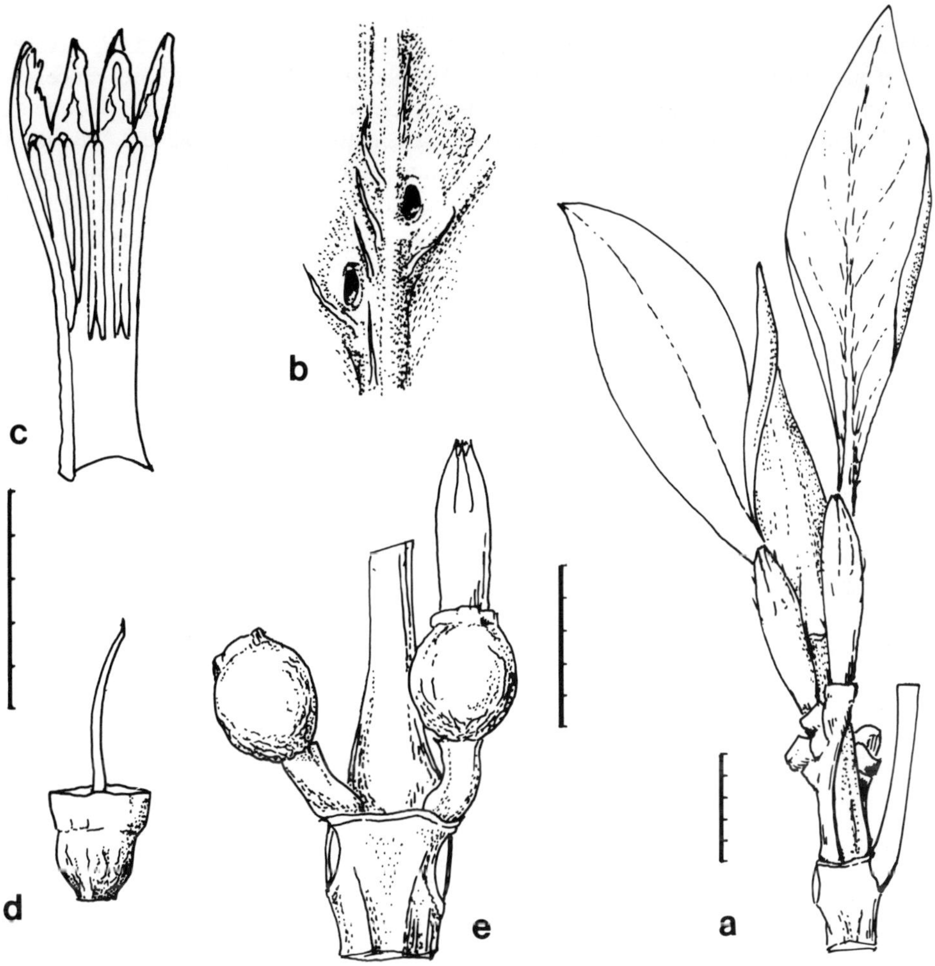

FIG. 8. *Timonius amungwiwanensis*. a. Tip of branchlet with staminate inflorescence. b. Magnified portion of abaxial leaf blade surface showing domatia in axils of secondary nerves. c. Corolla of staminate flower, opened to show stamens. d. Calyx of staminate flower with pistillodium. e. Node of branchlet with pair of pistillate flowers, one without corolla. Bars = 5 mm. (Based on: a–d, *Brass 4705*; e, *van Royen 11085*.)

prominulous above, elevated beneath, the secondary nerves obscure, 5–8 on each side of costa, spreading, the veinlets striate above and beneath, paxillate; domatia cryptlike and porate to pocketlike, appearing scattered over lamina beneath or distributed along costa, glabrous, often absent. Staminate inflorescences 2-branched, 9- to 19-flowered, at anthesis 2–4 cm long, 3–4 cm wide; peduncles 7–15 mm long, ca. 1 mm wide, strap-shaped, glabrous; bracts absent. Staminate flowers: calyx limb cupular, truncate to undulate, glabrous, the tube 2–3 mm long from base; corolla narrowly infundibular, white to greenish white, the tube to 11 mm long, to 2 mm wide, glabrous outside, glabrous within, the lobes 4, 2.4–3.5 mm long, 1.5–2 mm wide, ovate-elliptic, acute, with 3 prominent, subterete to subulate appendages extending from the subapical interior margin, these often quite unequal, to 2.5 mm long, the corolla limb ca. 5 mm in diameter, the lobes reflexed at maturity; stamens 4, the filaments very short, the anthers ca. 6.5 mm long, 0.5 mm wide, linear-clavate, glabrous; pistillodium 3–5 mm long, filiform, glabrous. Pistillate inflorescences 2-branched, (3-) 5- to 9-flowered, at anthesis to 3.5 cm long and 4 cm wide; peduncles 15–17 mm long, 1–1.5 mm wide, strap-shaped, glabrous; bracts absent. Pistillate flowers: calyx limb cupular, truncate to undulate, glabrous, the tube 0.5–1 mm long; corolla infundibular (seen in bud only, the tube ca. 2 mm wide), glabrous outside and within, the lobes 6, ovate, appendaged as in staminate corollas; staminodia 6, ca. 3.5 mm long, 0.2 mm wide, linear, glabrous; hypanthium ca. 2.5 mm long, ca. 2 mm in diameter, ellipsoid, glabrous, the disc glabrous, the style seen in bud only, irregularly 6-armed, glabrous. Fruits 4–5 mm long, 4.5–6 (–7) mm in diameter, subglobose, 5- or 6-costate, tuberculate, glabrous, color of mature fruits unknown; pyrenes 30–42 in 5–6 radiating double-files, subequal to unequal in length, obliquely oriented. Figs. 3g, 6n, o, 11c.

Phenology. Collected in flower between December and May, in fruit in March, July, August, and December.

Distribution. Philippines (Luzon, Mindoro, Biliran, Mindanao); in primary forests and ridge forests; 500–850 m.

Vernacular names. *Putútan, Upong-úpong* (Merrill, 1923, p. 541).

ADDITIONAL SPECIMENS EXAMINED. **Philippines**. LUZON. Quezon: Sierra Madre Mts., WSW of Baler, *Jacobs 7934* ♀ (A, PNH). Rizal: Mt. Lumutan, *Philipp. Bur. Sci. 29591* ♀ (A), *Philipp. Bur. Sci. 29778* ♀ (A). Camarines: without locality, *Philipp. Bur. Sci. 1505* ♀ (GH). Sorsogon: Mt. Bulusan, Irosin, *Elmer 15639* ♂ (A, GH), *Elmer 16636* ♀ (A, GH); without locality, *Philipp. Bur. Sci. 23330* ♀ (A).—BILIRAN. Mt. Suiro, N slope, *Philipp. Nat. Herb. 21621* ♂ (PNH).—MINDANAO. Surigao: Mt. Kabatuan, *Philipp. Nat. Herb. 10583* ♂ (A, PNH).

Timonius appendiculatus is a highly distinctive species similar to *T. pachyphyllus* (also from the Philippines) and differing from other species of subg. *Abbottia* in the prominent fingerlike appendages, which extend downward from the interior apices of staminate and pistillate corolla lobes; when the lobes are fully reflexed, the appendages stand suberect, forming a fimbriate corona surrounding the corolla throat. The species is also distinctive in the strongly 2-branched cymes with sessile, secund flowers. Multiflowered pistillate inflorescences are known in only a few other species of subg. *Abbottia*, e.g., *T. longifolius* and *T. zuckianus*, both from Papuasia. Merrill (1923) cites *T. appendiculatus* from Rizal, Batangas, Camarines, and Sorsogon Provinces of Luzon, and from Mindoro (Mt. Agong, *For. Bur. 9867*, n.v.); later collections, cited above, extend the species range to the islands of Biliran and Mindanao.

No type was designated among the three collections originally cited by Merrill, but a lectotype was indicated by Elmer in 1911. The type locality cited by Merrill is Bosoboso, but Elmer gives it as Antipolo, and this is what appears on the examined isolectotypes. The specimen "*B S 2161*," cited by Merrill in 1923, is almost certainly an error for *2164*.

4. Timonius avenis Valeton, Bull. Dép. Agric. Indes Néerl. 26: 46. 1909.—TYPE: [INDONESIA. IRIAN JAYA.] Western New Guinea, [Mimika Prov.] near Geluksheuvel, 15 Jul 1907, *Versteeg 1457* ♀ (holotype: BO!; isotypes: B! K! L! P! U!).

Epiphytic shrubs or climbers or free-standing trees to 40 m tall and 80 cm dbh. In var. *avenis*, bark reportedly smooth, pustular, pale gray to brown, the wood white to pale yellow; in var. *pubipetalus*, bark reportedly gray to brownish, rough, the wood dark red; nature of bark and wood unknown in vars. *magnifructus* and *vogelkopensis*. Branchlets 1–2 (–3) mm wide toward apex, pale gray to brown, glabrous to puberulent. Stipules to 50 mm long, 10 mm wide, long-lanceolate to ovate, acuminate, glabrous to minutely appressed-puberulent abaxially, glabrous adaxially. Petioles 3–10 (–15) mm long, ca. 1 mm wide at middle, semiterete to canaliculate to subterete, glabrous to minutely puberulent; leaf blades 1–11 cm long, 0.5–5 cm wide, elliptic to oblanceolate to obovate, sharply acute to acuminate or rarely obtuse at apex, acute to attenuate at base, chartaceous to coriaceous, glabrous to marginally puberulent above, glabrous to minutely appressed-puberulent especially on costa beneath, the costa plane to canaliculate above, elevated beneath, the secondary nerves indistinct, the veinlets prominulous above, submerged beneath, paxillate; domatia pitlike, scattered over lower surface of lamina and often also pocketlike along costa, or frequently absent. Staminate inflorescences 3- to 13-flowered, at anthesis 1–3 cm long, 1–3 cm wide; peduncles 1–7 (–12) mm long, 1–2 mm wide, compressed, glabrous to puberulent; bracts absent. Staminate flowers: calyx limb cupular, truncate to undulate to shallowly 4-dentate, sometimes rupturing with expansion of corolla, glabrous to scattered-puberulent, the tube 1–2 mm long from base; corolla slender-infundibular, the tube 7–10 mm long, ca. 1 mm wide, glabrous to appressed-puberulent outside, glabrous within, the lobes 4, 2–3 mm long, 0.5–1 mm wide, ovate to lanceolate, acute, papillose, the corolla limb 3–6 mm in diameter; stamens 4, the anthers to 6 mm long, 1 mm wide, linear, glabrous; pistillodium to 4 mm long, filiform, glabrous. Pistillate inflorescences 1-flowered; peduncles (1–) 3–15 mm long, 0.5–2 mm wide, compressed, glabrous. Pistillate flowers: calyx limb cupular, truncate to shallowly and irregularly denticulate, glabrous to scattered-puberulent, the tube 0.5–2 mm long; corolla infundibular to hypocrateriform, the tube 6–10 mm long, 1–2 mm wide, glabrous to appressed-puberulent outside, glabrous within, the lobes (4–) 6–10, 2–4 mm long, 0.5–1.5 mm wide, ovate to lanceolate to subulate, acute, papillose, the corolla limb to 8 mm in diameter; staminodia 6–8, 2–3 mm long, ca. 0.5 mm wide, linear, glabrous; hypanthium 1.5–3 mm long, 1–3 mm in diameter, subglobose to ovoid, glabrous, the disc glabrous, the style to 9 mm long, ca. 0.5 mm wide near middle, 6- to 10-armed, glabrous. Fruits 2–6 mm long, 4–9 mm in diameter, depressed-globose to short-cylindric, often somewhat laterally compressed, 5- to 8-costate when dry, glabrous, brownish purple when mature in var. *vogelkopensis*; pyrenes 30–75 in 6–8 radiating double-files, subequal, vertical. Figs. 3o, 6a, b.

Timonius avenis is most similar to *T. heptamerus* (Mt. Carstensz, western New Guinea), which has pale-strigillose corollas and somewhat larger ovaries (3–3.5 mm in diameter); otherwise, *T. heptamerus* could be accommodated in *T. avenis* var. *pubipetalus*. *Timonius trichanthus* (eastern New Guinea) and *T. dolichophyllus* (central New Guinea) differ from *T. avenis* in their (respectively) shaggy-tomentulose or densely pale-strigillose stipules versus the glabrous or at most appressed-puberulent stipules of *T. avenis*. *Timonius lamii* (western New Guinea) resembles *T. avenis* in its vertical and parallel pyrenes and glabrous vegetative parts, but is much more robust, with 3–4 mm wide branchlets and 10- to 17-flowered staminate inflorescences.

Timonius avenis, the most widely distributed species of subg. *Abbottia*, extends from eastern Borneo (Sambaliung) eastward to the southern highlands of Papua New Guinea, although the number of available collections is small (Fig. 9). Specimens assigned here show little variability in their vegetative features, except in degree of stipule pubescence and development of pocket-shaped domatia associated with the costa on leaf undersurfaces. Staminate inflorescences show considerable variability as to number of flowers and overall dimensions, but nearly the whole range of this variability can be seen in single collections when study material is ample enough. Variability in fruit size is roughly correlated with geographical distribution; specimens having especially large fruits occur in the Vogelkop region of western New Guinea and nearby Jappen Island. This character, as well as corolla pubescence, distribution of domatia, and shape of leaf blade apices, seem useful in distinguishing four taxa that are here considered worthy of varietal rank, and are distinguished in the following key.

KEY TO THE VARIETIES OF TIMONIUS AVENIS

1. Fruits 5–6 mm long, 7–9 mm in diameter; corolla of pistillate flowers glabrous; western New Guinea. 4b. *T. avenis* var. *magnifructus*.
1. Fruits 2–4 (–5) mm long, 4–5 (–7) mm in diameter; corolla of pistillate flowers glabrous or puberulent.
    2. Corollas and often also calyces gray appressed-pubescent; central New Guinea. 4c. *T. avenis* var. *pubipetalus*.
    2. Corollas and calyces usually glabrous or sparsely puberulent.
        3. Domatia small, open and pitlike, distributed over abaxial surfaces of leaf blades in axils of tertiary nerves; leaf blades sharply acute to acuminate at apex; stipules glabrous; southern New Guinea to Borneo. 4a. *T. avenis* var. *avenis*.
        3. Domatia pocketlike, distributed along abaxial costa and also often pitlike and scattered over lamina; leaf blades obtuse to acute at apex; stipules glabrous to minutely puberulent abaxially; Vogelkop Peninsula of New Guinea. 4d. *T. avenis* var. *vogelkopensis*.

4a. Timonius avenis var. **avenis.**

Stipules glabrous. Leaf blades sharply acute to acuminate at apex; domatia small, open, pitlike, distributed over lower leaf surfaces. Corollas of staminate and pistillate flowers glabrous to sparsely puberulent outside, the tube greenish to brownish, ca. 6 mm long (pistillate flowers), the lobes white. Fruits 2.5–4 mm long, 4–5 mm in diameter; color of mature fruits unknown. Fig. 3o.

Phenology. Collected in flower in January, June, July, and August, in fruit in March, June, July, August, and November.

Distribution (Fig. 9). Scattered localities in eastern Borneo, the Moluccas, and New Guinea; secondary growth or riverine swamp forests; 40–1130 m.

Vernacular Names (from specimen labels). *Ramdoeatene* (Fakfak); *Seranai* (Kutei, Borneo); *Tagraea* (Menubar, Borneo).

ADDITIONAL SPECIMENS EXAMINED. **Indonesia**. KALIMANTAN. [*Borneo.*] KUTEI: Menubar (Sangkulirang), *bb 34694* (BO, L), *Kostermans 5229* ♂ (A, BO, CANB, L, LAE); Susuk region, *Kostermans 5711* ♂ (BO, L).—[*Moluccas.*] OBI: Anggai, Gunung Batu Putih, *de Vogel 4122* ♀ (BO, L, LAE).—CERAM: Sosokoetai, *Eyma 2552* ♂ (BO, L).—IRIAN JAYA [*Western New Guinea*]: Fakfak, Genofa Mtn, *BW 10275* ♀ (A, CANB, L, LAE, SING). **Papua New Guinea.** [*Papua.*] WESTERN: Yat, *NGF 31978* ♀ (CANB, K, L, LAE); Angarmaruk, *NGF 33417* ♀ (CANB, K, L, LAE).—GULF: above Purari River, 63 km from Baimuru, *LAE 61164* ♀ (LAE).

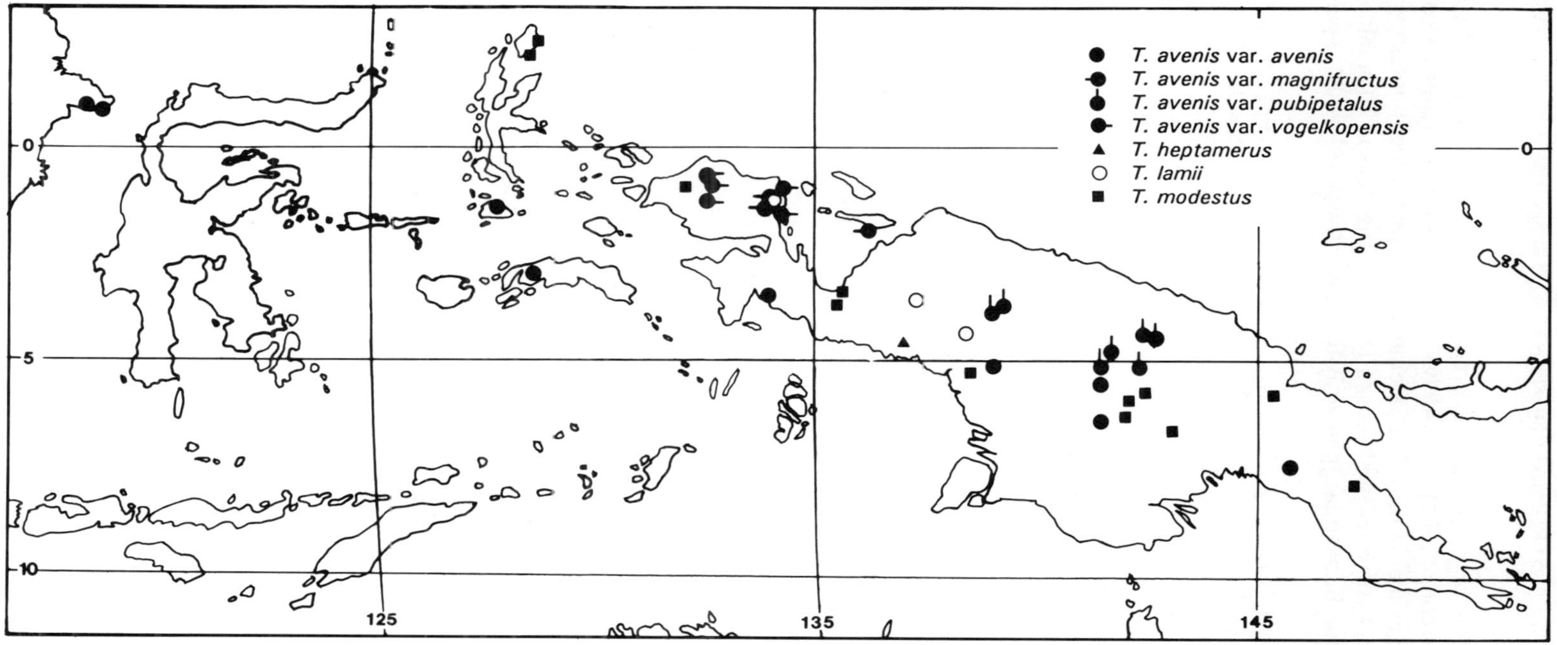

FIG. 9. Distribution of *Timonius avenis* var. *avenis*, *T. avenis* var. *magnifructus*, *T. avenis* var. *pubipetalus*, *T. avenis* var. *vogelkopensis*, *T. heptamerus*, *T. lamii*, and *T. modestus*.

The delineation of *T. avenis* var. *avenis* is not at all satisfactory, because it is based on very few collections from widely scattered localities; staminate collections, for example, are available only from Ceram and Borneo. Field notes characterize the variety as epiphytic trees or shrubs or free-standing trees to 40 m tall and 80 cm dbh, the bark smooth, pustular, pale gray to brown, the wood white to pale yellow.

4b. Timonius avenis var. **magnifructus** S. Darwin, var. nov.—TYPE: [INDONESIA. IRIAN JAYA.] Western New Guinea. Vogelkop Peninsula: Arfak, Anggi Gita Lake, 1800 m, 9–22 Oct 1948, *Kostermans 2220* ♀ (holotype: BO!; isotypes: A! K! L! SING!).

Differt haec varietas ab affinibus fructibus magnis (5–6 mm longis 7–9 mm latis) et tubo corollae pistillatae longiore (usque ad 10 mm).

Stipules glabrous to finely appressed-puberulent toward abaxial base. Leaf blades sharply acuminate at apex; domatia pocketlike, distributed along abaxial costa, or domatia absent. Corollas of staminate and pistillate flowers glabrous, the tube 8–10 mm long, reddish, the lobes white. Fruits 5–6 mm long, 7–9 mm in diameter; color of mature fruits unknown. Fig. 6a, b.

Phenology. Collected in flower and fruit in January, April, and October.

Distribution (Fig. 9). Vogelkop region of western New Guinea and Jappen Island; common at edges of forests; 1800–2300 m.

ADDITIONAL SPECIMENS EXAMINED. **Indonesia.** IRIAN JAYA [*Western New Guinea*]. Vogelkop Peninsula: Anggi, Arfak Mts., *Kanehira & Hatusima 13608* ♀ (A); Mt. Koreimat above Testega, Anggi Lakes, *BW 14130* ♀ (BISH, BRI, L, LAE); Arfak, Anggi Gita Lake, *Kostermans 2265* ♂ (BO).—JAPPEN ISLAND: Geelvink Bay, Senèmi near Serui, *Aët & Idjan 314* ♀ (L).

Variety *magnifructus* is perhaps confined to high elevations of the Arfak Mountains of New Guinea. It differs from other varieties of *T. avenis* in its somewhat larger fruits, but is approached in that respect by var. *vogelkopensis*. The Jappen Island specimen cited above, with immature fruits, is tentatively assigned here.

4c. Timonius avenis var. **pubipetalus** Valeton, Bot. Jahrb. Syst. 61: 37. 1927 ('pubipetala'). *Timonius pubipetalus* (Valeton) Merrill & Perry, J. Arnold Arb. 26: 236. 1945.—TYPE: [PAPUA NEW GUINEA.] Northeastern New Guinea. East Sepik: "Sepikgebiet" [Felsspitze], Aug 1913, *Ledermann 12678* ♀ (lectotype, here designated: L!; isolectotype: LAE!).

Stipules gray appressed-pubescent to glabrous. Leaf blades sharply acuminate at apex; domatia small, open, pitlike, distributed over lower leaf surfaces. Corollas of staminate and pistillate flowers appressed-puberulent outside, the tube 7–10 mm long, pinkish, the lobes white. Fruits 3–4 mm long, 4–5 mm in diameter; color of mature fruits unknown.

Phenology. Flowers have been observed in February, March, May, and October.

Distribution (Fig. 9). Taritatu River [Idenburg River] of western New Guinea eastward to Sepik region; primary or secondary rain forest; 850–2000 m.

Vernacular names. *Omai* (S Highlands); *Tirok* (Oksapmin).

ADDITIONAL SPECIMENS EXAMINED. **Indonesia.** IRIAN JAYA [*Western New Guinea*]. Djajapura: 6 km SW of Bernhard Camp, Idenburg River, *Brass & Versteegh 12592* ♂ (A, BM, BO, BRI, L, LAE); 4 km SW of Bernhard Camp, *Brass 13687* ♂ (A, BM, BO, BRI, K, L, LAE); Bernhard Camp, *Brass 12878* ♂ (A, BM, BO, BRI, L, LAE). **Papua New Guinea** [*Northeastern New Guinea.*] WEST SEPIK: Mogofogola River, upstream from junction with Bielga River on track from Folongonom to Busilmin, *LAE 59583* ♂ (BISH, K, L, LAE, M); Oksapmin, *NGF 41515* ♂ (A, BISH, BRI, K, L, LAE, SING), *NGF 41516* ♀ (A, BISH, BRI, CANB, K, L, LAE, SING), *NGF 41766* ♀ (A, BRI, CANB, K, L, LAE).—EAST SEPIK: Felsspitze, *Ledermann 12870* ♀ (L, LAE).—[*Papua.*] WESTERN: Ok Tedi headwaters near Kennecott camp, *NGF 42807* ♀ (A, BISH, BRI, CANB, K, L, LAE).—SOUTHERN HIGHLANDS: Anderbia near Tagari footbridge, *UPNG 1725* ♀ (L, LAE).

Timonius avenis var. *pubipetalus* is similar to var. *avenis*, but differs in the frequently puberulent branchlets, calyces, and corollas; the leaves are like those of *T. secundiflorus*, but lack prominent pocketlike domatia. Available collections were taken from epiphytic shrubs or forest trees to 22 m tall and ca. 40 cm dbh with gray to brownish, rough bark and dark red wood.

Three collections, *Ledermann 12809*, *12678*, and *11370*, were cited by Valeton in the protologue of *T. avenis* var. *pubipetalus*, but only the second collection, *Ledermann 12678*, appears to be extant; it is represented in the Rijksherbarium, Leiden, by a sheet bearing only the locality "Sepikgebiet" and includes a pistillate specimen of *T. avenis* var. *pubipetalus* as well as a staminate specimen of *T. secundiflorus*. For the collection Valeton provides additional data, presumably taken from the Berlin specimen: "Blüten rötlich-grün" and the locality "Felsspitze, 1400 m." There is nothing to indicate that Valeton did not consider both elements of *Ledermann 12678* as representing *T. avenis* var. *pubipetalus*; the pistillate element is here designated the lectotype of Valeton's variety. *Timonius avenis* var. *pubipetalus* is also represented by *Ledermann 12870*, from the same locality but not cited by Valeton.

A few collections from central New Guinea (e.g., *LAE 59583, NGF 41515, NGF 41516, NGF 41766, NGF 42807, UPGN 1725*), listed above, resemble *T. avenis* var. *pubipetalus* except for more coarsely pubescent stipules and undersurfaces of leaf blades. With respect to indument and domatia, they approach *T. trichanthus*, which has an adjacent, eastwardly extending geographical range. Until a more extensive suite of collections is available, they are accommodated in the present taxon.

4d. Timonius avenis var. **vogelkopensis** S. Darwin, var. nov.—TYPE: [INDONESIA. IRIAN JAYA.] Western New Guinea. Vogelkop Peninsula: Tanah Merah, N side of Lake Ajamaru, 220 m, 10 Mar 1962, *BW 15269* ♀ (holotype: A!; isotypes: B! BRI! CANB! K! L! LAE! Z!).

Idem ac varietas avenis sed domatiis glabris vel puberulis marsupiformibus in folii lamina abaxiali prope costam prominentibus, et stipularum partibus exterioribus, foliorum laminarum extremitatibus, et inflorescentiis saepe puberulis differt.

Stipules glabrous to minutely puberulent. Leaf blades obtuse to acute at apex; domatia pocketlike, distributed along abaxial costa, and also often open and pitlike, distributed over lower leaf surfaces. Corollas of staminate and pistillate flowers thinly puberulent to glabrous outside, the tube ca. 6 mm long in pistillate flowers, to 10 mm long in staminate flowers, brownish to pinkish, the lobes white. Fruits 2–4 (–5) mm long, 5–7 mm in diameter, brownish purple when mature.

Phenology. Collected in flower and fruit in March, May, and November.

Distribution (Fig. 9). Vogelkop Peninsula of western New Guinea; secondary or primary and often riverine forests on limestone; 30–840 m.

Vernacular names. *Arsekum*, *Wassekoem*, *Wissekum* (Maibrat).

ADDITIONAL SPECIMENS EXAMINED. **Indonesia.** IRIAN JAYA [*Western New Guinea*]. Vogelkop Peninsula: central part of Tamrau Range, S slope, from Sudjak to Mt. Kusemun, Aiwa River, *van Royen & Sleumer 7730* ♀ (A, CANB, K, L), *van Royen & Sleumer 7737* ♂ (A, CANB, K, L, LAE); Tabun, N of Lake Ajamaroe, *BW 15353* ♂ (A, B, CANB, K, L, LAE, Z); Ajamaroe, *BW 4975* ♀ (CANB, L, LAE); Beri Creek, near Andai, SW of Manokwari, *BW 11936* ♀ (BISH, L, LAE); Waren, 60 mi S of Manokwari, *Kanehira & Hatusima 13141* ♂ (A).

The few available collections were taken from shrubs or climbers becoming trees to 30 m tall and 70 cm dbh. Merging in fruit dimensions with *T. avenis* var. *magnifructus*, also from the Vogelkop Peninsula, var. *vogelkopensis* perhaps represents a series of populations from generally lower elevations.

5. Timonius belensis Merrill & Perry, J. Arnold Arb. 26: 236, fig. 2A. 1945.—TYPE: [INDONESIA. IRIAN JAYA.] Western New Guinea. Bele River, 18 km NE of Lake Habbema, common in secondary forest, 2400 m, Nov 1938, *Brass 11527* ♀ (holotype: A!; isotypes: BM! BRI! L! LAE!).

Epiphytic shrubs or commonly understory trees or canopy trees to 30 m tall and 50 cm dbh with large and spreading crowns. Bark gray to brown, pustular or scaly and peeling, the lenticels prominent, the underbark pinkish, the wood white to straw-colored and hard. Branchlets (3–) 4–5 mm wide toward apex, dark to pale reddish brown, glabrous but often ciliate above stipule scars. Stipules to 50 mm long and 8 mm wide, lanceolate, acuminate, glabrous abaxially, glabrous adaxially except for colleters near base. Petioles to 25 mm long, (1–) 2–3 mm wide at middle, semi-to subterete, glabrous; leaf blades (3–) 6–16 cm long, (1–) 3–6 cm wide, elliptic to obovate to suborbicular, broadly acute to briefly acuminate at apex, attenuate to rounded at base, thick-coriaceous, glabrous or with scattered hairs near apex above, glabrous beneath, the costa plane to canaliculate above, elevated beneath, the secondary nerves hardly discernible, spreading, the veinlets obscure to prominulous above and beneath, paxillate; domatia pocket- or pitlike along costa and often also scattered over abaxial lamina, glabrous or with a few marginal hairs. Staminate inflorescences (1-) 3- to 7-flowered, at anthesis to 1.5 cm long and 1.5–2 cm wide; peduncles 2–8 (–14) mm long, 1–2 mm wide, somewhat compressed, glabrous; bracts absent. Staminate flowers: calyx limb cupular, truncate to undulate, glabrous, the tube 1–2 mm long from base; corolla tubular-infundibular, white or greenish, the tube 6–7 mm long, 2 mm wide, glabrous or with scattered hairs outside, glabrous within, the lobes 4, to 3.5 mm long and 3 mm wide, ovate, obtuse, papillose, the corolla limb to 9 mm in diameter; stamens 4, anthers not seen; pistillodium 5–6 mm long, slender-clavate, glabrous. Pistillate inflorescences 1-flowered; peduncles 2–18 mm long, 2–2.5 mm wide, somewhat compressed, glabrous; bracts absent. Pistillate flowers: calyx limb cupular-cylindric, truncate, glabrous, the tube 1–1.5 mm long; corolla tubular-infundibular, white to greenish, the tube 2–6 mm long, 2.5–3 mm wide, glabrous outside, glabrous within, the lobes (6–) 8–12, 2.5–3.5 mm long, 0.5–1.5 mm wide, linear-lanceolate, acute, papillose, the corolla limb 5–8 mm in diameter; staminodia ca. 10, ca. 2.5 mm long, linear-sagittate, glabrous; hypanthium 3.5–5 mm long, 4–5 mm in diameter, subglobose, glabrous, the disc glabrous, the style 5–7 mm long, ca. 2 mm wide near middle, 6- to 10-armed, glabrous. Fruits 6–8 mm long, 8–11 mm in diameter, subglobose, often laterally compressed, tuberculate or ribbed, glabrous, black when mature; pyrenes 40–80, usually in 4 radiating double-files, subequal, obliquely to subobliquely oriented. Fig. 7a.

Phenology. Flowering and fruiting throughout the year.

Distribution (Fig. 13). Vicinity of Lake Habbema (western New Guinea), and highlands of eastern New Guinea; lower primary montane forests to subalpine forests, moss-forests, and ridgetops; 80–3300 m, but usually above 2000 m.

Vernacular names (from specimen labels). *Kumai* (Togoba), *Omai* (Mendi), *Homai* (S. Highlands), or *Kuma* (W. Highlands); *Abalan* (Wabag); *Haikapula* (Ialibu); *Mesgukeys* (Madang); *Mogon* (W. Sepik); *Tognap* (Nondugl); *Tomburump* (Minj); *Tsubaga* (Kutubu).

ADDITIONAL SPECIMENS EXAMINED. **Indonesia.** IRIAN JAYA [*Western New Guinea*]. Djajapura: Bele River, 18 km NE of Lake Habbema, *Brass & Versteegh 11110* ♀ (A, BO, BRI, L); Mt. Trikora (Mt. Wilhelmina), *Mangen 356* ♀ (L). **Papua New Guinea.** [*Northeastern New Guinea.*] WEST SEPIK: Mt. Entaldam, S of Busilmin airstrip, *LAE 67013* ♀ (A, BISH, LAE).—MADANG: N Kaironk Valley, *Bulmer 134* ♀ (LAE).—MOROBE: Budemu, *NGF 21272* ♀ (A); 10.5 km NE of Wau, *Pratt NG-78–1184* (LAE).—WESTERN HIGHLANDS: Tambul-Tomba rd, Nebelyer Divide, *Robbins 463* ♂ & ♀ (CANB, LAE); Porget logging area, 12 mi N of Wabag, *NGF 11285* ♂ (A, CANB, K, L, LAE); Minj-Nona Divide, N slopes Kubor Range, *Pullen 5216* ♂ (A, CANB, K, L, LAE); Alipe, Kepaka, upper Kaugel, *Bowers 376* ♀ (CANB); Malapuku, Kulumindi, *Bowers 528* ♂ (CANB, K, LAE).—EASTERN HIGHLANDS: Mt. Wilhelm, E slopes, *Brass 30219* ♀ (A, CANB, K, L, LAE), *Brass 30346* ♀ (A, CANB, K, L, LAE), *Brass 30619* ♂ (A, L, LAE); Kombugomambumo, Keglsugl–Pindaunde trail, *ANU 7400* (LAE); 8 km ENE of Kundiawa, *Paijmans 1257* ♀ (CANB, LAE); Kuaki River above Taromambuno Mission, *Pullen 294* ♀ (A, BM, BRI, CANB, L, LAE); above Fatima River, *NGF 40730* ♀ (A, BISH, BRI, CANB, E, K, L, LAE, SING); Marafunga, *NGF 24614* ♀ (A, BO, BRI, CANB, LAE), *LAE 51091* ♂ (BRI, CANB, L, LAE); Asaro–Mairifutica Divide, 3 mi NW of Daulo camp, *Pullen 437* ♀ (A, CANB, L, LAE); Kini Creek, NE slopes Mt. Michael, *NGF 11375* ♀ (A, CANB, L, LAE), *NGF 11727* ♂ (A, BRI, CANB, LAE); upper Omahaiga Valley, *Pullen 589* ♀ (CANB, L, LAE); Mt. Otto, S slopes, *Brass 30837* ♀ (A, CANB, K, L, LAE), *Brass 31056* ♀ (CANB, K, L, LAE), *Brass 31093* ♀ (A, BO, CANB, K, L, LAE); Kortumi S.D.A. Mill via Goroka, *NGF 6751* ♂ (A, BM, BO, CANB, K, L, LAE); Kainantu-Okapa rd near Onamuga, *Hartley 13687* ♀ (A, BRI, CANB, L, LAE).—[*Papua.*] SOUTHERN HIGHLANDS: Mt. Ambua, *Kalkman 5040* ♀ (A, C, CANB, L, LAE, SING), *Vink 17461* ♀ (A, BISH, CANB, L, LAE, MO, RSA); near Lei Camp, SE slope Mt. Ambua, *NGF 26986* ♀ (A, BISH, BRI, CANB, E, K, L, LAE, M, SING); Ibiwara, Tari Gap, *NGF 25119* ♀ (A, BISH, BRI, CANB, K, L, LAE, SING), *NGF 25188* ♀ (A, BISH, BRI, CANB, K, L, LAE); Habona rest house, 6.5 mi W of Mt. Ne, *NGF 28497* ♀ (A, BISH, BRI, CANB, K, L, LAE, SING); N slopes Mt. Kerewa, *Kalkman 4731* ♂ (A, C, CANB, L, LAE, SING); near Imu, Lake Kutubu, *Schodde 2228* ♀ (A, BRI, CANB, K, L, LAE); near Capuchin Mission Sawmill, N side Pangia Rd, 4 mi E of Ialibu, *UPNG 4224* ♀ (A, LAE, UPNG); ca. 3 mi N Pundia, S slopes Mt. Giluwe, *Saunders 1051* (CANB, L, LAE); Iaro logging area, S slopes Mt. Giluwe, *LAE 63162* ♀ (A, BISH, L, LAE, M, QRS).—WESTERN: 30 km NE of Tabubil, *Hyndman 300* ♀ (UPNG).—Not mapped: Eastern Highlands: ca. 5 mi W of Keglsugl airstrip, *Saunders 772* ♂ (A, BM, BRI, CANB, L, LAE); Western Highlands: Autu Mountain, 5 mi NNE Nandugl, *Simonett 126* (LAE).

As here interpreted, *T. belensis* is one of the most frequently collected species of subg. *Abbottia*, distinguishable by its relatively robust vegetative parts, which are usually glabrous or at most sparsely appressed-puberulent, and the stout corolla tubes (of staminate and pistillate flowers) more than 2 mm wide. The leaf blades are thick-coriaceous with prominulous venation above and beneath; the secondary nerves are usually distinguishable from the veinlets but are not prominent. In most fruits, the pyrenes radiate obliquely, but in others the pyrenes are nearly vertical, suggesting *T. pubistipulus*. The type and a few other collections from the vicinity of Mt. Wilhelmina in western New Guinea are well out of range for the species; perhaps additional collecting will reveal populations of *T. belensis* in the intervening region.

A few collections from scattered localities in eastern New Guinea have exceptionally large leaf blades (to 20 cm long and 10 cm wide) lacking domatia, and large stipules (to 10 cm long): Western Highlands, Yaibos, near Wabag, *ANU 2323* (CANB, L, LAE); Eastern Highlands, Mt. Wilhelm, Komanemambuno, *ANU 10623* (CANB); upper Omahaiga valley, *Pullen 560* (CANB); ca. 0.5 mi NW of Daulo, *Saunders 841* (CANB, LAE); Gulf Prov., Wabo area, *LAE 66351* (LAE). All such specimens are sterile (although notes sometimes refer to separate fruits, collected from ground, or flowers) and probably all represent coppice shoots. Two collections from the eastern edge of the species range (*NGF 21272*, and *Pratt 78–1184*, cited above) have narrower and smoother leaf blades suggesting *T. bismarckensis* but are otherwise like *T. belensis*. A collection from Mt. Bosavi, Western Province of Papua (*Jacobs 8894* ♀, BISH, L, LAE) resembles *T. modestus* vegetatively, but is otherwise assignable to *T. belensis*.

Timonius belensis is frequently associated with *Nothofagus*, *Castanopsis*, and *Podocarpus*. Fruits are reported as being eaten in quantity by forest birds.

6. Timonius bismarckensis S. Darwin, sp. nov.—Type: Papau New Guinea. [*Bismarck Archipelago.*] New Britain: 10 mi S of Cape Ruge, Talasea, limestone pothole region, ridgetop rain forest, 180 m, 11 May 1966, *NGF 26763* ♀ (holotype: A!; isotypes: BM! CANB! L! LAE! SING!).

Timonius ex affinitate T. neobritannici S. Darwin sed foliis majoribus crassicoriaceis plerumque ellipticis sine domatiis prominentibus et pedunculis post anthesin longioribus ligulatis differt.

Epiphytic shrubs becoming buttressed free-standing trees to 27 m tall and 60 cm dbh. Bark rough, scaly, fissured, or with prominent lenticels, the underbark brownish to orange, the wood white to pale straw-colored. Branchlets 2–3 mm wide toward apex, pale, glabrous or rarely puberulent. Stipules to 45 mm long and 10 mm wide, lanceolate, acuminate, glabrous or sparsely appressed-puberulent abaxially, glabrous or with scattered, appressed hairs adaxially. Petioles 5–15 mm long, 1–2 mm wide at middle, narrowly canaliculate, glabrous; leaf blades 3–12 (–16) cm long, 1.5–6 cm wide, elliptic to somewhat obovate, sharply acute to acuminate at apex, acute to attenuate at base, stiff-chartaceous to coriaceous, glabrous above, glabrous beneath, the costa plane to canaliculate above, elevated beneath, the secondary nerves indistinguishable, the veinlets prominulous, paxillate; domatia absent or rarely present as small, open pits on lamina beneath, glabrous. Staminate inflorescences not seen. Pistillate inflorescences 1-flowered; peduncles (6–) 10–15 mm long, 1.5–2 mm wide, compressed, glabrous; bracts absent. Pistillate flowers: calyx limb annular-cupular, truncate, glabrous, the tube 0.5 mm long; corolla seen in bud only, white, the tube to 6 mm long, 1.5–2.5 mm wide, glabrous or minutely puberulent outside, the lobes 6–8, 2–3 mm long, 0.5–1 mm wide, lanceolate, acute, papillose, the corolla limb 5–6 mm in diameter; staminodia not seen; hypanthium 3–4 mm long, 4–5 mm in diameter, subglobose, glabrous, the disc glabrous, the style 8-armed, glabrous. Fruits 5–6 mm long, 7–8 mm in diameter, depressed-globose, shallowly ribbed, glabrous, maroon when mature; pyrenes ca. 36 and perhaps in 8 radiating double-files, subequal, obliquely oriented. Figs. 3f, 10.

Phenology. Probably flowering and fruiting throughout the year.

Distribution (Fig. 16). Bismarck Archipelago (New Britain, New Ireland, and Manus Island); frequent in hillside forests on limestone; 300–800 m.

ADDITIONAL SPECIMENS EXAMINED. **Papua New Guinea.** [*Bismarck Archipelago.*] NEW BRITAIN: 7 mi SE of Benim, *NGF 27355* ♀ (A, CANB, L, LAE), *NGF 27372* ♀ (A, CANB, K, L, LAE); near Linga Linga, *NGF 49482* ♀ (A, K, L, LAE); Mt. Mataleloch, *LAE 51169* ♀ (A, CANB, K, LAE); Lackit, *NGF 38024* ♀ (L, LAE).—NEW IRELAND: 5 mi S of Logagon, N Schleinitz Range, *LAE 65579* ♀ (A, BISH, BM, L, LAE, M, MO, QRS, SING).—MANUS ISLAND: Wylli Valley, W of Mt. Dremsel, E of Watani River, *Sands 954* ♀ (LAE).

Timonius bismarckensis is distinguishable from *T. neobritannicus* in its larger, more elliptic, and more thickly coriaceous leaf blades, which lack prominent domatia; the fruits are borne on longer and more conspicuously strap-shaped peduncles. As here interpreted, *T. bismarckensis* is confined to land areas surrounding the Bismarck Sea. A few collections from eastern New Guinea (e.g., *NGF 21272* and *Pratt NG 78–1184*) approach *T. bismarckensis* in having narrow leaf blades with the secondary nerves hardly distinguishable and small (sometimes absent) domatia, but they otherwise resemble *T. belensis* (no. 5); however, the occurrence of *T. bismarckensis* in eastern New Guinea is not unexpected.

7. **Timonius bougainvillensis** Merrill & Perry, J. Arnold Arb. 26: 249, fig. 8B. 1945.—TYPE: SOLOMON ISLANDS. Bougainville Island: Kupei Gold Field, common in rain forest, 900 m, 4 Oct 1930, *Kajewski 1677* ♀ (holotype: A!; isotypes: BISH! BM! BO! BRI! G! L! P! S! SING!).

Epiphytic shrubs to slender or spreading trees to 15 m tall. Bark smooth, sometimes with small nodules, gray to brown, the underbark pale brown, the wood cream-brown to white. Branchlets to 1.5 mm wide toward apex, dark red-brown, glabrous or with scattered hairs. Stipules to 45 mm long and 7 mm wide, lanceolate, long-acuminate, glabrous abaxially, glabrous adaxially except for colleters near base. Petioles 8–25 mm long, to 1 mm wide at middle, semiterete, glabrous; leaf blades (3–) 6–13 cm long, (1–) 2–4 cm wide, oblong to lanceolate or elliptic, sharply acuminate at apex, broadly acute to rounded to abruptly acuminate at base, coriaceous to chartaceous, glabrous above, glabrous or with scattered, stramineous, appressed to spreading hairs ca. 2 mm long on costa and occasionally also on lamina beneath, the costa plane to canaliculate above, elevated beneath, the secondary nerves 11–13 (–15) on each side of costa, spreading, the veinlets prominulous to plane, paxillate; domatia pit- to pocketlike or cryptlike and porate in axils of secondary nerves and often also scattered over lamina beneath, often absent, glabrous. Staminate inflorescences (3-) 5- (7-)flowered, at anthesis to 10 mm long and 13 mm wide, congested; peduncles to 6 mm long, ca. 1 mm wide, glabrous;

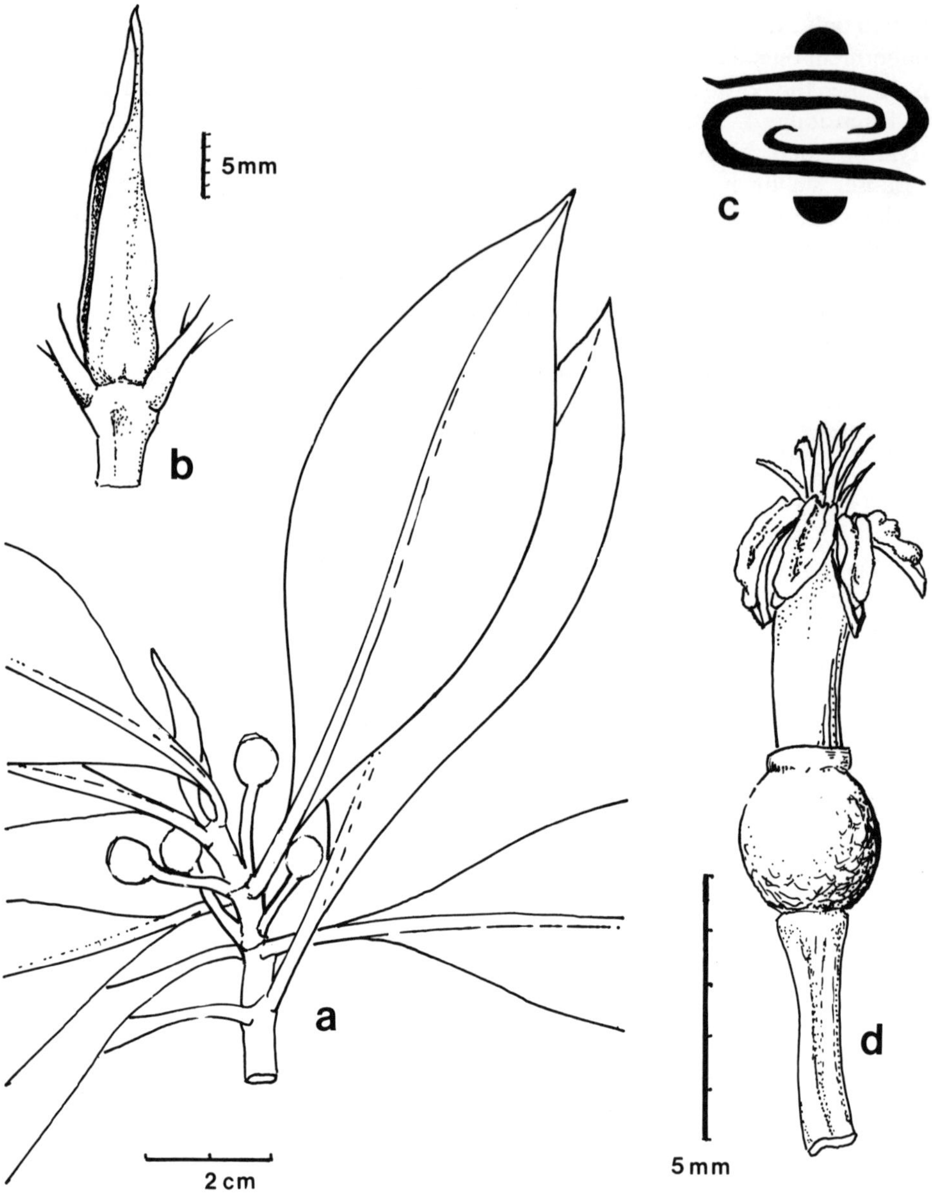

FIG. 10. *Timonius bismarckensis*. a. Tip of branchlet with fruits. b. Terminal bud showing imbricate stipules. c. Cross section (diagramatic) through terminal bud showing stipule imbrication. d. Pistillate flower. (Based on: a, *LAE 51169*; b–d, *NGF 27355*.)

bracts absent or obscure, to 1 mm long, glabrous. Staminate flowers: calyx limb urceolate-cylindric, truncate to undulate, minutely scattered-puberulent, the tube ca. 4.5 mm long from base; corolla seen in bud only, white, the tube to 6.5 mm long, ca. 2.5 mm wide, subglabrous to densely strigillose on lobes outside, glabrous within, the lobes 4, to 5 mm long in late bud, lanceolate-ovate; stamens 4, the filaments to 1 mm long, the anthers 5–5.5 mm long, ca. 0.8 mm wide, linear-oblong, glabrous; pistillodium to 5.5 mm long, filiform, glabrous. Pistillate inflorescences 1-flowered; peduncles to 12 mm long, to 1 mm wide, glabrous; bracts absent or obscure, to 0.5 mm long. Pistillate flowers: calyx limb cylindric-cupular, truncate to undulate, glabrous, the tube 1.5–2 mm long; corolla infundibular, white, the tube 4–7 mm long, to 4 mm wide, glabrous outside, glabrous to minutely strigillose around throat within, the lobes 6, 2–3 mm long, 2–3 mm wide, broadly ovate, erosulous, papillose, the corolla limb 6–9 mm in diameter; staminodia 6, 3–4 mm long, ca. 0.8 mm wide, oblong-sagittate, glabrous; hypanthium to 4 mm long, to 4 mm in diameter, subglobose, glabrous, the disc glabrous, the style to 8 mm long, ca. 1 mm wide near middle, irregularly 6-armed, glabrous. Fruits (4.5–) 6–8 (–9) mm long, 5–6 (–11) mm in diameter, subglobose, tuberculate when dry, glabrous, black when mature; pyrenes 90–200, subequal, oblique. Figs. 3e, 7g, h, 11a, b.

Phenology. Flowering and fruiting July to October.

Distribution (Fig. 12). Solomon Islands (Bougainville, Kolombangara, and Guadalcanal Islands); common in rain forests and stunted cloud forests; 760–1525 m.

Vernacular Name. *Kentorku* (Buin).

Additional Specimens Examined. **Solomon Islands.** Bougainville: Kupei Gold Field, *Kajewski 1750* ♀ (A, BISH, BM, BO, BRI, G, P); S rim of Lake Loloru crater, *Schodde & Craven 3911* ♀ (A, BRI, CANB, K, L, LAE), *Schodde & Craven 3914* ♂ (A, BRI, CANB, K, L, LAE); S upper slopes Lake Loloru crater, *Schodde & Craven 3745* ♀ (A, B, BISH, BRI, CANB, K, L, LAE); Lake Loloru, *NGF 31299* ♂ & ♀ (A, BRI, CANB, L, LAE), *NGF 31382* ♂ (A, BRI, CANB, L, LAE, SING); mts near Bokarkani, *NGF 31572* ♂ (A, BRI, CANB, L, LAE); Koniguru, Buin, *Kajewski 2049* ♀ (A, BISH, BO, BRI, C, G, L, P, S, SING).—Kolombangara: upper campsite, *Roy. Soc. Exped. 2493* ♀ (A, K, L, LAE); Kolombangara, *Roy. Soc. Exped. 1209* ♀ (A, K, L, LAE).—Guadalcanal: Mt. Gallego, E ridge, *Roy. Soc. Exped. 6066* ♀ (A, K, L, LAE, P, SING); Popomanasiu, *Roy. Soc. Exped. 150* ♀ (K, L).

Timonius bougainvillensis, apparently confined to the islands of Bougainville, Kolombangara, and Guadalcanal, is unusual among species of subg. *Abbottia* in having leaf blades with well-developed lateral nerves (and a usually thin-chartaceous lamina); the finer reticuli-paxillate venation is typical of the subgenus. Bracts, although quite small, are nonetheless discernible in the pistillate and congested staminate inflorescences, whereas inflorescence bracts are generally absent in other species of the subgenus. The bracts and lateral nerves may be ancestral features retained in this geographically isolated species, although the solitary pistillate flowers are probably derived from cymose inflorescences, as found in *T. longifolius* and *T. zuckianus* from New Guinea.

8. Timonius carstensensis Wernham in Ridley, Trans. Linn. Soc. London, Bot. 9: 73. 1916.—Type: [Indonesia. Irian Jaya.] Western New Guinea. Utakwa River to Mt. Carstensz, Camp x, xi, 27 Jan 1913, *Boden Kloss* ♀ (lectotype, here designated: BM!).

Small trees? Branchlets 2–2.5 mm wide toward apex, pale gray or brown, glabrous. Stipules to 35 mm long and 15 mm wide, lanceolate, acuminate, glabrous abaxially, glabrous adaxially except for colleters at base. Petioles 5–10 mm long, ca. 2 mm wide at middle, glabrous; leaf blades 5–11 cm long, 1.5–4 cm wide,

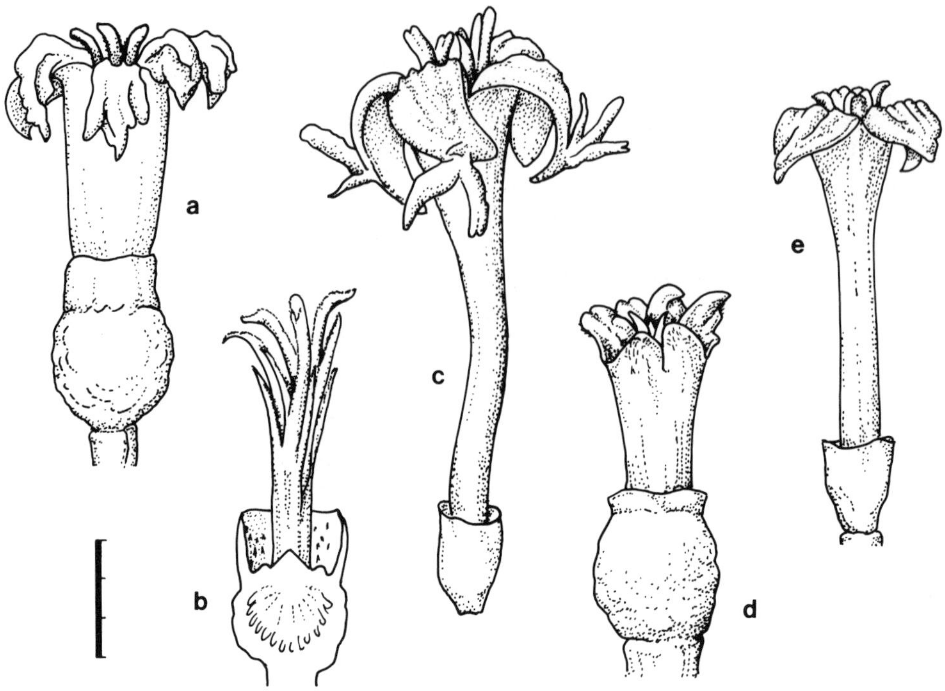

FIG. 11. *Timonius bougainvillensis*, *T. appendiculatus*, and *T. carstensensis*. a, b. *T. bougainvillensis*, pistillate flower. a. Exterior view. b. Vertical section showing style and glandular interior of calyx limb. c. *T. appendiculatus*, staminate flower. d, e. *T. carstensensis*. d. Pistillate flower. e. Staminate flower. Bar = 3 mm. (Based on: a, b, *NGF 31299*; c, *PNH 10583*; d, e, *Boden Kloss s.n.*)

elliptic, acuminate at apex, acute at base, thick-coriaceous, glabrous above, glabrous or with scattered hairs on costa beneath, the costa plane to canaliculate above, elevated beneath, the secondary nerves indistinct, the veinlets submerged in lamina, paxillate; domatia elongate, cryptlike and porate, distributed along costa beneath, sparsely pubescent. Staminate inflorescences congested, 5-flowered, at anthesis ca. 1 cm long and 1.5 cm wide; peduncles 1–2 mm long, ca. 3 mm wide, compressed, glabrous; bracts absent. Staminate flowers: calyx limb cupular, truncate to undulate, glabrous, the tube 2–3 mm long from base; corolla narrowly infundibular, the tube to 8 mm long, ca. 1 mm wide, glabrous outside, glabrous within, the lobes 4, 2–2.5 mm long, ca. 2 mm wide, broadly ovate, acute, papillose, the corolla limb to 6 mm in diameter; stamens 4, the anthers ca. 4 mm long, 1 mm wide, oblong-elliptic, subsagittate at base, glabrous; pistillodium not seen. Pistillate inflorescences 1-flowered; peduncles 10–12 mm long, ca. 1.5 mm wide, glabrous; bracts absent or minute and rounded, to 0.5 mm long. Pistillate flowers: calyx limb shallowly cupular, undulate to 7- or 8-lobed, glabrous, the tube ca. 1 mm long, the lobes ca. 0.5 mm long, ca. 1.5 mm wide; corolla infundibular, the tube 3.5–5 mm long, 2–3 mm wide, glabrous to finely puberulent distally outside, glabrous within, the lobes 6, 2–2.5 mm long, 1–1.5 mm wide, lanceolate, acute, papillose, the corolla limb to 5 mm in diameter; staminodia 6, to 2.5 mm long and 1 mm wide, linear-oblong, glabrous; hypanthium ca. 3 mm long, subglobose, glabrous, the disc

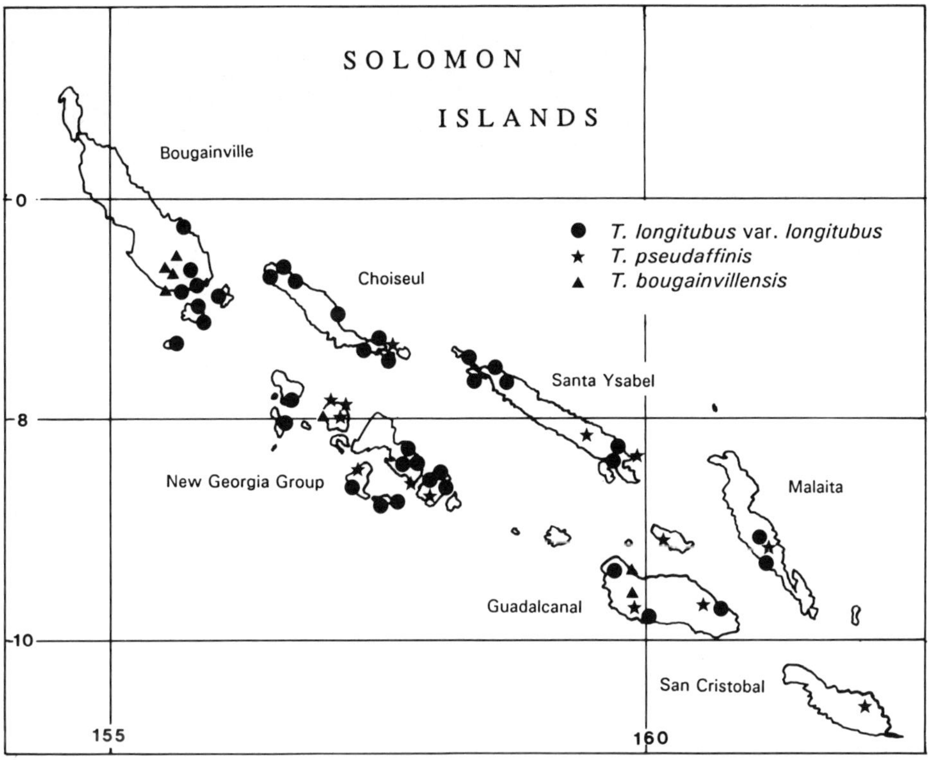

FIG. 12. Distribution of *Timonius longitubus* var. *longitubus*, *T. pseudaffinis*, and *T. bougainvillensis*.

glabrous, the style ca. 6 mm long, ca. 1 mm wide near middle, glabrous. Fruits ca. 7 mm long, ca. 7 mm in diameter, subglobose, glabrous, color of mature fruits unknown; pyrenes ca. 30, subequal, vertical. Figs. 4f, 11d, e.

Distribution (Fig. 13). Western New Guinea, vicinity of Mt. Carstensz; scrub forests and open ridgetops; 2000–3350 m.

ADDITIONAL SPECIMEN EXAMINED. **Indonesia.** IRIAN JAYA [*Western New Guinea*]. Snow Mts: Utakwa River to Mt. Carstensz, camp xi, xii, *Boden Kloss* ♂ (BM).

Timonius carstensensis, still known only from the two original collections, is unusual in having subglobose (neither compressed nor prominently ribbed) fruits crowned by a subterminal crenate-undulate calyx, and a prominent, glabrous disc (this reminiscent of *T. lamii*). The available specimens are otherwise referable to *T. trichocladus*, also known from west-central New Guinea, but in that species the pyrenes are not vertically arranged in fruit. Until more material is available, *T. carstensensis* is retained as a separate species, but resembling *T. avenis*.

9. Timonius cryptophlebus S. Moore in C. T. White, Proc. Roy. Soc. Queensland 34: 55. 1923.—TYPE: [PAPUA NEW GUINEA.] Papua. Central: Dilava, Jul/Aug 1918, *C. T. White 428* ♀ (holotype: BM!; isotype: BRI!).

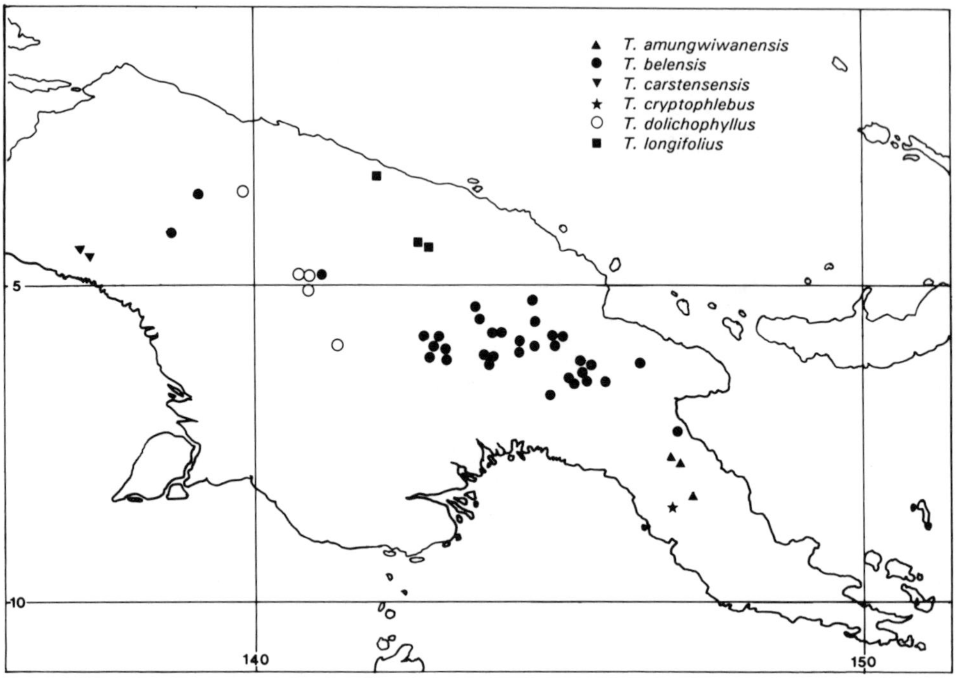

FIG. 13. Distribution of *Timonius amungwiwanensis*, *T. belensis*, *T. carstensensis*, *T. cryptophlebus*, *T. dolichophyllus*, and *T. longifolius*.

Small trees; nature of bark and wood unknown. Branchlets 3.5–4.5 mm wide toward apex, reddish brown, densely pubescent with reddish hairs to 0.5 mm long. Stipules to 17 mm long and 3 mm wide, lanceolate, acuminate, densely strigillose with white to yellow hairs to 1 mm long abaxially, canescent adaxially. Petioles 1–12 mm long, ca. 1.5 mm wide at middle, semiterete, puberulent; leaf blades 8–11 cm long, 4–4.5 cm wide, elliptic, bluntly acute at apex, attenuate to acute to abruptly truncate at base, coriaceous, scattered-puberulent above, subvelutinous with spreading, reddish hairs to 1 mm long scattered over lamina beneath, the costa plane to canaliculate above, elevated beneath, the secondary nerves indistinct, the veinlets prominulous, paxillate; domatia absent. Staminate inflorescences not seen. Pistillate inflorescences 1-flowered. Pistillate flowers: calyx limb cupular-cylindric, truncate, glabrous, the tube 1–2 mm long; corolla narrowly infundibular, the tube 10–11 mm long, 2–3 mm wide, densely white- to yellow-strigillose outside, glabrous within, the lobes 7(?); staminodia 7, 3–3.5 mm long, ca. 0.7 mm wide, linear-oblong, glabrous; hypanthium ca. 2.5 mm long, ca. 3 mm in diameter, depressed-cylindric, glabrous, the disc glabrous, the style to 9 mm long, to 1.5 mm wide near middle, 6- to 8-armed, glabrous. Fruits 5–6 mm long, ca. 8 mm in diameter, depressed-cylindric, 6- to 8-costate, glabrous, color of mature fruits unknown; pyrenes ca. 30 and probably arranged in 6 or 8 double-files, subequal, vertical.

Timonius cryptophlebus, as yet known only from the type (locality shown in Fig. 13), is most similar to *T. avenis* and its relatives (fruits with pyrenes subequal and vertically oriented). It is perhaps conspecific with *T. trichanthus*, which is not known from Papua, but extends into southern Morobe Province of northeastern

New Guinea. *Timonius cryptophlebus* differs from *T. trichanthus* in the complete absence of leaf domatia and the densely reddish pubescence on branchlets and leaf blades. Without intermediate collections, the two taxa are here provisionally maintained as separate species.

10. Timonius dolichophyllus (Merrill & Perry) S. Darwin, stat. et comb. nov. *Timonius trichanthus* var. *dolichophyllus* Merrill & Perry, J. Arnold Arb. 26: 239, fig. 3D. 1945.—TYPE: [INDONESIA. IRIAN JAYA.] Western New Guinea. Vicinity of Idenburg River, banks of streams, 15 km SW of Bernhard Camp, 1500 m, Jan 1939, *Brass 12398* ♀ (holotype, A!; isotypes: BM! BO! BRI! L! LAE!).

Epiphytic shrubs eventually becoming trees to 6 m tall; nature of bark and wood unknown. Branchlets to 4 mm wide toward apex, gray to reddish brown, glabrous to spreading- or appressed-puberulent. Stipules to 40 (rarely to 65) mm long and 8 mm wide, lanceolate, acute, densely white- to yellow-strigillose abaxially with hairs to 1 mm long, glabrous adaxially except for colleters near base. Petioles 4–18 mm long, 1.5–2 mm wide at middle, stiff, semiterete, spreading-puberulent; leaf blades 4–18 cm long, 1–5 cm wide, long-elliptic to subfalcate, sharply acuminate at apex, acute to attenuate at base, subcoriaceous to chartaceous, glabrous above except frequently puberulent on costa toward base, spreading- to appressed-strigillose over lamina beneath with stramineous hairs ca. 1 mm long, the costa canaliculate above, elevated beneath, the secondary nerves indistinguishable, the veinlets prominulous above, paxillate; domatia open and pitlike, scattered over lamina and often also along costa beneath, frequently absent, glabrous. Staminate inflorescences (5-) 7- to 11-flowered, at anthesis to 2.5 cm long and 3 cm wide; peduncles 3–7 mm long, to 1.5 mm wide, pubescent; bracts absent. Staminate flowers: calyx limb cupular-cylindric, truncate to shallowly 4-dentate, sparsely to densely pubescent, the tube to 2.5 mm long from base; corolla narrowly infundibular, greenish to white, the tube to 10 mm long, ca. 1 mm wide, densely strigillose outside with stramineous hairs ca. 0.5 mm long, glabrous within, the lobes 4, to 3 mm long and 1 mm wide, ovate-deltate, acute, papillose, the corolla limb 5–6 mm in diameter; stamens 4, the filaments short, the anthers ca. 3.5 mm long, linear, glabrous; pistillodium not seen. Pistillate inflorescences 1-flowered; peduncles to 9 mm long, ca. 1 mm wide, spreading-pubescent to subglabrous; bracts absent. Pistillate flowers: calyx limb cupular, truncate, puberulent to subglabrous, the tube ca. 1 mm long; corolla infundibular, greenish to white, the tube (3–) 6–8 mm long, 1.5–2 mm wide, densely stramineous-strigillose outside, glabrous within, the lobes 6, to 3 mm long and 1 mm wide, ovate, acute, papillose, the corolla limb to 7 mm in diameter; staminodia 6, ca. 3 mm long, linear-sagittate, glabrous; hypanthium 3–3.5 mm long, 3–4 mm in diameter, subglobose, glabrous except toward apex, the disc glabrous, the style to 9 mm long, 6-armed, glabrous. Fruits 3.5–6 mm long, 5–7.5 mm in diameter, subglobose, 6- to 8-costate when dry, glabrous, purplish when mature; pyrenes 40–60 in 6–8 radiating double-files, subequal, vertical.

Phenology. Collected in flower and fruit in January, and May to July.

Distribution (Fig. 13). Central New Guinea (Star Mountains to Fly River); in primary or secondary forests; 80–1500 m.

ADDITIONAL SPECIMENS EXAMINED. **Indonesia.** IRIAN JAYA [*Western New Guinea*]. Star Mountains: below Okdenam, Oksibul River, *Soegeng Reksodihardjo 532* ♀ (BO, K, L), *Soegeng Reksodihardjo 547* ♂ (K, L), *Soegeng Reksodihardjo 579* ♂ (K, L); Okwalimkam headwaters, *NGF 33179* ♂ (CANB, K, L, LAE), *NGF 33115* ♀ (A, BRI, CANB, K, L, LAE, SING); Mt. Antares, Bon River near mouth of Minam, *Kalkman 4337* ♀ (A, BM, BO, CANB, L, LAE, SING); Mt. Antares, 1 km E of junction of Bon & Minam Rivers, *Kalkman 4384* ♀ (A, BM, CANB, L, LAE). **Papua New Guinea.** [*Papua*.] WESTERN: Fly River, 528 mile camp, *Brass 6804* ♀ (A, BM, BO, BRI, L, LAE).

Timonius dolichophyllus is one of the most easily recognizable species of subg. *Abbottia*, especially in the yellowish to white-strigillose stipules, long-lanceolate and often subfalcate leaf blades, and strongly 6- to 8-lobed fruits. The species is most similar to *T. trichanthus*, where Merrill and Perry had assigned it varietal status, but that species differs in having shaggy-tomentulose stipules and broadly elliptic to oblanceolate leaf blades.

11. Timonius heptamerus Wernham in H. Ridley, Trans. Linn. Soc. London, Bot. 9: 73. 1916.—TYPE: [INDONESIA. IRIAN JAYA.] Western New Guinea. Utakwa River to Mt. Carstensz, camp iii, ca. 760 m, Feb 1913, *Boden Kloss* ♀ (holotype: BM!).

Small trees? Branchlets to 2 mm wide toward apex, grayish, puberulent. Stipules to 42 mm long and 7 mm wide, lanceolate, acuminate, scattered appressed-puberulent abaxially. Petioles 10–15 mm long, 1–1.5 mm wide at middle, semi-terete, glabrous to puberulent; leaf blades 6–9 cm long, 1.5–3 cm wide, elliptic, sharply acute to acuminate at apex, attenuate at base, thinly coriaceous, glabrous above, pubescent with scattered, appressed, stramineous hairs beneath, the costa plane to canaliculate above, elevated beneath, the secondary nerves indistinct, the veinlets submerged in lamina, paxillate; domatia cryptlike and porate, distributed along costa beneath, pubescent. Staminate inflorescences not seen. Pistillate inflorescences 1-flowered; peduncles 4–6 mm long, 1–2 mm wide, puberulent; bracts absent. Pistillate flowers: calyx limb cupular, truncate, subglabrous, the tube ca. 1 mm long; corolla infundibular, the tube 4–5 mm long, ca. 2 mm wide, densely pale-strigillose outside, glabrous within, the lobes 6–8, 2–2.5 mm long, ca. 1 mm wide, lanceolate to ovate, acute, papillose, the corolla limb to 7 mm in diameter; staminodia 6–8, ca. 2 mm long, 0.3 mm wide, linear, glabrous; hypanthium ca. 2 mm long, 3–3.5 mm in diameter, subglobose to depressed-cylindric, subglabrous, the style to 6.5 mm long, ca. 0.5 mm wide near middle, 6- to 8-armed, glabrous. Fruits (immature) ca. 5 mm long, ca. 5 mm in diameter; pyrenes ca. 45 in 8 radiating double-files, subequal, vertical.

Timonius heptamerus, known only from the type (locality shown in Fig. 9), seems most similar to *T. avenis* var. *pubipetalus*, which occurs from the Sepik region westward into Djajapura province (Idenburg River). *Timonius heptamerus* differs in having somewhat larger leaf blades, longer and broader petioles, domatia distributed alongside the costa beneath, somewhat larger stipules, completely truncate calyx limbs of pistillate flowers, and stramineous-strigillose (rather than grayish puberulent) corollas. Further collecting may demonstrate that *T. heptamerus* should be accommodated within *T. avenis*, already a wide-ranging and polymorphic species with locally distinguishable populations.

12. Timonius lamii S. Darwin, sp. nov.—TYPE: INDONESIA. IRIAN JAYA. [Western New Guinea.] Doormantop ridge, in moss-forest, 1450 m, 9 Oct 1920, *Lam 1572* ♂ (holotype: L!; isotypes: A! BO! BRI! L! SING!).

Species T. modesti Merrill & Perry aspectu similis, sed stipulis glabris et pyrenis verticalibus differt.

Scrambling shrubs or trees to 5 m tall and 10 cm dbh; nature of bark and wood unknown. Branchlets 3–4 mm wide toward apex, glabrous and often glaucous. Stipules to 35 mm long and 7 mm wide, long-oblong-lanceolate, acuminate, glabrous abaxially, glabrous adaxially except for colleters near base and sometimes a few hairs at apex. Petioles 10–20 mm long, 1–2 mm wide at middle, stiff, semi- to subterete, glabrous; leaf blades (6–) 8–14 cm long, (2–) 3–5.5 cm wide, elliptic to oblanceolate, sharply acuminate at apex, long-attenuate at base, stiff-coriaceous, glabrous above, glabrous except over domatia beneath, the costa plane to canaliculate above, elevated and rounded beneath, the secondary nerves hardly discernible, the veinlets submerged beneath, weakly striate above, paxillate; domatia cryptlike, porate or the ostiole slitlike, pubescent, distributed along costa beneath, also glabrous pitlike or cryptlike domatia distributed over lamina, but either type occasionally absent. Staminate inflorescences 11- to 17-flowered, at anthesis to 1.5 cm long and 3 cm wide; peduncles 4–6 mm long, 1–2 mm wide, compressed, glabrous; bracts absent. Staminate flowers: calyx limb cupular-cylindric, truncate, glabrous, reddish, the tube to 3 mm long from base; corolla seen in bud only, white or greenish, the tube 8 mm long, ca. 1 mm wide, glabrous outside, glabrous within, the lobes 4, 3–4 mm long, lanceolate, acute; stamens 4, the filaments short, the anthers ca. 3 mm long, 0.3 mm wide, oblong-sagittate, glabrous; pistillodium ca. 3 mm long, unbranched. Pistillate inflorescences 1-flowered; peduncles 5–12 mm long, ca. 2 mm wide, stiff, compressed, glabrous; bracts absent. Pistillate flowers: calyx limb cupular-cylindric, truncate to undulate, glabrous, reddish, the tube 1–2 mm long; corolla seen in bud only, white, glabrous outside, glabrous within, the lobes 7–8; staminodia 7–8, glabrous; hypanthium 2–3 mm long, 3–4 mm in diameter, subglobose to obovoid, glabrous, the disc glabrous, in fruit becoming as high as calyx limb with prominent, pale style scar at apex, the style not seen. Fruits 6–7 mm long, 7–9 mm in diameter, depressed-cylindric, 6- or 7-sulcate, glabrous, color of mature fruits unknown; pyrenes ca. 50 in 8 radiating double-files, subequal, vertical. Fig. 14.

Phenology. Collected in flower in June, August, and October, in fruit in August.

Distribution (Fig. 9). Scattered localities in western New Guinea; primary forests; 1450–2000 m.

ADDITIONAL SPECIMENS EXAMINED. **Indonesia.** IRIAN JAYA [*Western New Guinea*]: Vogelkop Peninsula, Arfak Mountains, Beribai, *BW 11462* ♂ (A, CANB, L, LAE); Snow Mountains, Baliem Valley above Wellesey, *Kostermans & Soegeng 767* ♀ (A, L).

Timonius lamii resembles *T. modestus* in its large, psilate leaf blades, but stipules in the latter species are pubescent and the pyrenes are obliquely oriented in the fruits. At present, *T. lamii* is known from three widely separated localities at relatively high elevations in western New Guinea. Additional high-altitude collec-

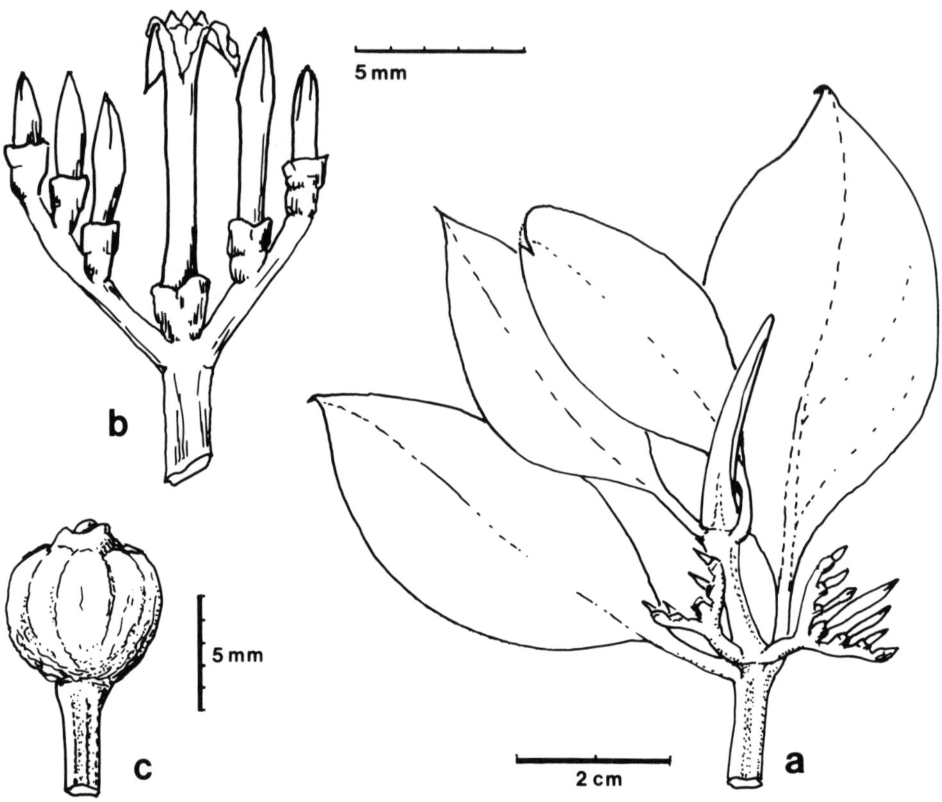

FIG. 14. *Timonius lamii.* a. Tip of branchlet with pair of staminate inflorescences. b. Staminate inflorescence. c. Fruit. (Based on: a, b, *Vink 11462*; c, *Kostermans & Soegeng 767.*)

tions may confirm a close relationship with *T. wollastonii*, in which the corolla lobes are more broadly proportioned and the disc is much less conspicuous in fruit.

13. Timonius longifolius Valeton, Bot. Jahrb. Syst. 61: 44. 1927.—SYNTYPES: PAPUA NEW GUINEA. East Sepik: Hunsteinspitze, 200–400 m, 10 Aug 1912, *Ledermann 8212*, *8213a* (B, destroyed).—TYPE: PAPUA NEW GUINEA. East Sepik: Yapa (Hunstein) River, straggling shrub ca. 5 m tall on river bank, 135 m, 26 Jul 1966, *Hoogland & Craven 10663* ♀ (neotype, here designated: A!; isoneotypes: BRI! CANB! K! L! LAE).

Straggling shrubs, climbers, or trees to 8 m tall; nature of bark and wood unknown. Branchlets 2–3 mm wide toward apex, brownish, puberulent but soon glabrous. Stipules to 60 mm long and 8 mm wide, lanceolate, acuminate, densely stramineous-strigillose to brown-spreading-puberulent to subglabrous abaxially, glabrous adaxially except strigose at very base below colleters. Petioles 6–18 mm long, ca. 1 mm wide at middle, semi- to subterete, spreading brown-puberulent; leaf blades oblanceolate, 6–19 cm long, 2–5.5 cm wide, sharply acute to acuminate at apex, acute to abruptly rounded at base, chartaceous, glabrous above except puberulent on costa toward base, glabrous to scabridulous with scattered, suberect hairs over lamina or pubescent only on costa beneath, the costa impressed above, elevated beneath, the secondary nerves usually hardly distinguishable, spreading from costa, the veinlets prominulous above, paxillate; domatia small tufts of hairs along costa beneath or absent. Staminate inflorescences 16- to 25-flowered, at anthesis to 5 cm long and 4 cm wide; peduncles 10–14 mm long, 1–1.5 mm wide, compressed, scattered-puberulent; bracts absent. Staminate flowers: calyx limb cupular, truncate to undulate, minutely strigillose outside, glabrous within, the tube 1–2 mm long from base; corolla seen in bud only, white, the tube ca. 5 mm long, ca. 1.5 mm wide, finely puberulent outside, the lobes 4, to 1.5 mm long in bud; stamens not seen, pistillodium not seen. Pistillate inflorescences 1- to 5-flowered, at anthesis to 5 cm long and 5 cm wide; peduncles 8–25 mm long, 1–2 mm wide, compressed, scattered spreading-puberulent with brownish hairs. Pistillate flowers: calyx limb cupular, truncate to undulate, scattered-puberulent outside, glabrous within, the tube ca. 1 mm long; corolla seen in bud only, white, glabrous to puberulent on lobes outside, the lobes probably 6; staminodia not seen; hypanthium subglobose, glabrous, the disc glabrous, the style not seen. Fruits 3–3.5 mm long, 5–6.5 mm in diameter, depressed-cylindric, 5- or 6-costate, glabrous, purple-brown when mature; pyrenes 40–60 in 6 radiating double-files, subequal, vertical. Figs. 3c, 4d, e, 6c, d, 15d.

Phenology. Collected in flower and fruit in July and August.

Distribution (Fig. 13). A component of riverbank vegetation in the Sepik River region of New Guinea; 130–200 m.

Vernacular names. *Shipren* (Waskuk); *Tindili* (Orne); *Yagutu* (Wagu).

ADDITIONAL SPECIMENS EXAMINED. **Papua New Guinea.** [*Northeastern New Guinea.*] WEST SEPIK: near Wantipi, Bliri River, *Darbyshire & Hoogland 8355* ♀ (A, BRI, CANB, G, K, L, LAE).—EAST SEPIK: Yapa (Hunstein) River, *Hoogland & Craven 10747* ♂ (CANB, K, L, LAE).

Timonius longifolius resembles *T. zuckianus* in having pistillate flowers arranged in cymes, which is perhaps a primitive character state versus the solitary pistillate flowers (reduced inflorescences) characterizing subg. *Abbottia*. In pubescence and stipule features, *T. longifolius* is most similar to *T. dolichophyllus*.

Valeton cited two collections, *Ledermann 8212* and *8213a*, in the protologue of *T. longifolius*. Neither collection is now at B (they were presumably destroyed) nor have duplicates been located elsewhere. The application of Valeton's name is clear from the original description, and the neotype is from the region visited by Ledermann.

14. Timonius longitubus Merrill & Perry, J. Arnold Arb. 26: 241, fig. 4B. 1945.—TYPE: SOLOMON ISLANDS. Santa Ysabel Island: Tiratona, tree 15 m tall in rain forest at 600 m, 8 Dec 1932, *Brass 3316* ♀ (holotype: A!; isotypes: BISH! BM! BO! BRI! L!).

Epiphytic shrubs, "stranglers" or eventually free-standing trees to 30 m tall. In var. *longitubus*, bark reportedly smooth to pustular or fissured, brownish to grayish, the wood white to pale brown or yellowish; nature of bark and wood unknown in var.

pubescens. Branchlets 4–6 mm wide toward apex, pale gray to brown, densely stramineous-strigillose, at length glabrous; stipules to 100 mm long and 22 mm wide, lanceolate to broadly elliptic, acuminate, appressed-strigillose or rarely spreading-pubescent abaxially, densely strigose to subglabrate-ciliate adaxially. Petioles 18–40 mm long, 1.5–2.5 mm wide at middle, stiff, semiterete, glabrous; leaf blades (5–) 8–16 (–20) cm long, (2.5–) 4–8 (–9) cm wide, obovate to broadly elliptic, abruptly sharp-acuminate to cuspidate at apex, attenuate at base, coriaceous, glabrous or rarely puberulent above, glabrous to scattered-puberulent or rarely densely velutinous beneath, the costa prominulous above, elevated beneath, the secondary nerves prominulous above and beneath, 5–7 on each side of costa, spreading, the veinlets prominulous above, paxillate; domatia cryptlike and porate along costa and often also scattered over lamina beneath, sometimes absent, glabrous to puberulent. Staminate inflorescences 3- to 5- (9-)flowered, at anthesis 2–5 cm long, 2–3 cm wide; peduncles 12–30 mm long, 1.5–2.5 mm wide, glabrous; bracts absent or obscure. Staminate flowers: calyx limb cupular, truncate to undulate, often ruptured by expanding corolla, glabrous or sometimes ciliate, the tube 3–10 mm long from base; corolla narrowly infundibular, the tube 18–22 mm long, 1.5–3 mm wide, glabrous outside, glabrous within, the lobes 4, 7–15 mm long, ca. 5 mm wide, broadly ovate, apiculate, papillose, the corolla limb to 30 mm in diameter; stamens 4, the filaments ca. 1 mm long, the anthers 5–8 mm long, 0.5–1 mm wide, linear-oblong, glabrous; pistillodium 5–7 mm long, filiform, glabrous. Pistillate inflorescences 1-flowered; peduncles 12–30 mm long, 1.5–2.5 (–4) mm wide, glabrous to pubescent; bracts absent or obscure. Pistillate flowers: calyx limb cupular, truncate to undulate, frequently ruptured by expanding corolla, glabrous to ciliate, the tube 3–6 mm long; corolla infundibular, the tube 10–13 mm long, 2.5–5 mm wide, usually glabrous outside, glabrous within, the lobes 6–8, 5–12 mm long, 2–5 mm wide, lanceolate, apiculate, papillose, the corolla limb to 11 mm in diameter; staminodia 6–8, 3.5–4 mm long, 0.5–1 mm wide, subsagittate, glabrous; hypanthium 4–7 mm long, 4–6 mm in diameter, subglobose, usually glabrous, the style to 18 mm long, ca. 1.5 mm wide near middle, irregularly 10-armed, glabrous. Fruits (12–) 15–18 (–25) mm long, 13–18 (–25) mm in diameter, subglobose, tuberculate-costate, glabrous, reddish when mature in var. *longitubus*; pyrenes 400–500 (or more ?) arranged around a central cavity, subequal. Figs. 3a, 7d, 15b, c.

Timonius longitubus is a distinctive species with exceptionally stout branchlets and relatively large flowers. Secondary nerves of leaf blades are prominulous, as in *T. bougainvillensis*, but in that species they number more than 10 on each side of costa and the blades are oblong in outline. *Timonius longitubus* is well collected in the Solomon Islands; a single collection from New Britain is referable to this species, but it differs in being conspicuously pubescent and is here considered to represent a distinct variety.

Key to the Varieties of Timonius longitubus

1. Leaf blades glabrous above, glabrous or rarely with scattered hairs on costa and nerves beneath; stipules abaxially pubescent with appressed hairs; Solomon Islands. 14a. *T. longitubus* var. *longitubus*.
1. Leaf blades persistently pubescent above with short and spreading hairs, densely pubescent (velutinous) with spreading hairs over lamina beneath; stipules abaxially densely spreading-pubescent; New Britain. 14b. *T. longitubus* var. *pubescens*.

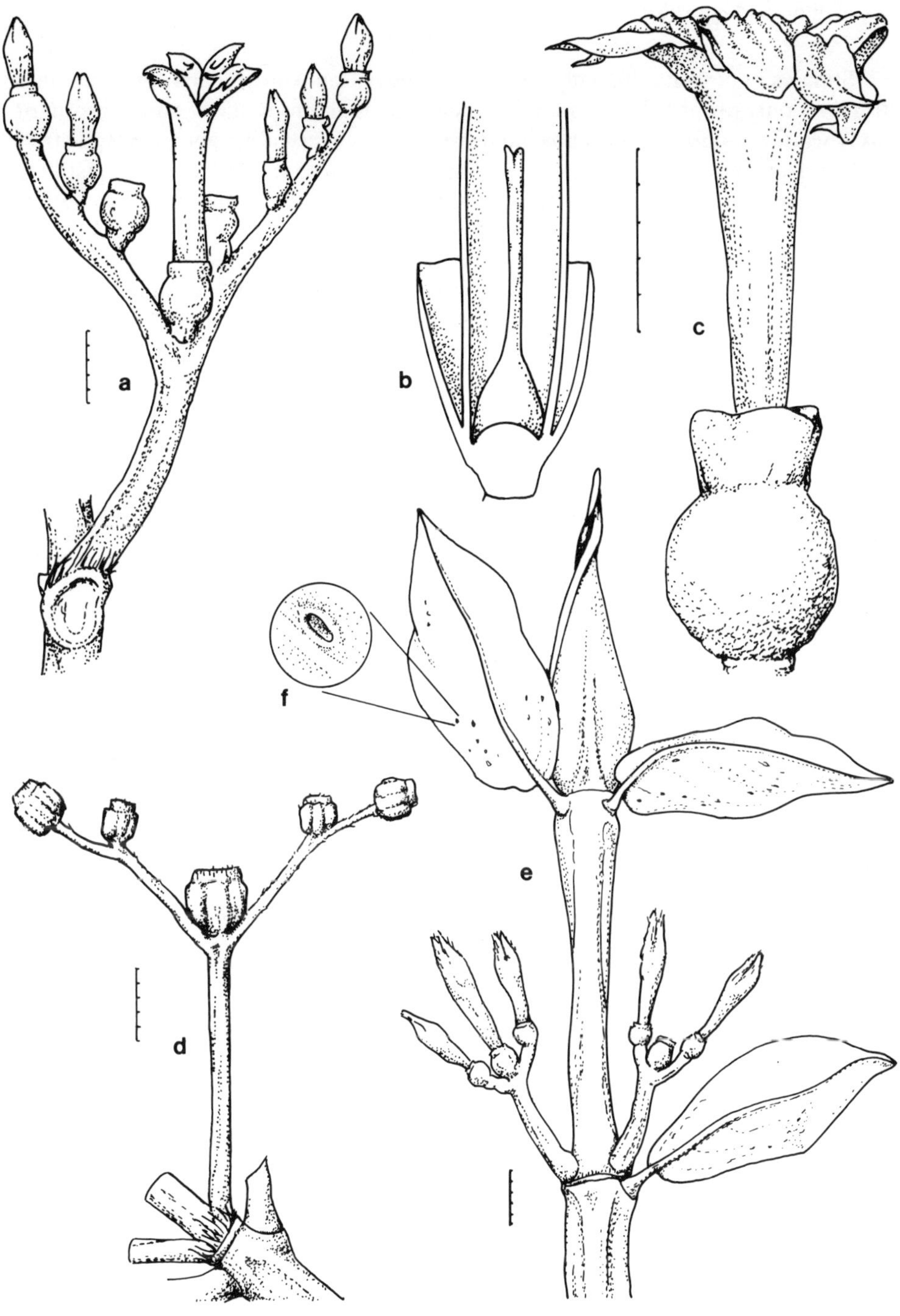

FIG. 15. *Timonius pachyphyllus*, *T. longitubus*, *T. longifolius*, and *T. oktediensis*. a. *T. pachyphyllus*, staminate inflorescence. b, c. *T. longitubus*. b. Vertical section of base of staminate flower showing pistillodium. c. Pistillate flower. d. *T. longifolius*, pistillate inflorescences with fruits. e, f. *T. oktediensis*. e. Distal end of branchlet showing terminal, imbricate stipules and a pair of staminate inflorescences. f. Magnified portion of abaxial leaf blade surface showing domatium. Bars = 5 mm. (Based on: a, *BS 42277*; b, *Craven 39980*; c, *Roy. Soc. Exped. 221*; d, *Hoogland & Craven 10663*; e, f, *NGF 42793*.)

14a. Timonius longitubus var. **longitubus.**

Stipules pubescent abaxially with appressed, often whitish hairs, or sometimes strigillose only toward abaxial margins. Leaf blades glabrous above and beneath, or rarely scattered-pubescent on abaxial nerves; domatia cryptlike and porate, distributed on leaf undersurfaces near costa, sometimes absent. Peduncles of pistillate inflorescences 1.5–2 mm wide toward middle, glabrous. Flowers white to yellow and faintly scented. Fruits reddish when mature.

Phenology. Flowering and fruiting throughout the year.

Distribution (Fig. 12). Solomon Islands from Bougainville to Guadalcanal; in primary and secondary forests on swampy to well-drained soils; 0–1250 m.

Vernacular names. *Simidi, Simindi, Aisimidi, Semende,* or *Siwidi* (Kwara'ae); *Boli* or *Bula* (Kwara'ae); *Balaho* (Santa Ysabel); *Uguroto* (Buin); *Tehor* (Vulolo, Guadalcanal).

ADDITIONAL SPECIMENS EXAMINED. **Solomon Islands.** BOUGAINVILLE: Pavairi, *NGF 30650* ♀ (A, BISH, BRI, CANB, K, L, LAE, M, SING); Crown Prince Mts, *NGF (Corner) s.n.* (LAE); lower slopes Lake Loloru crater, *Schodde & Craven 3864* ♀ (A, BRI, CANB, K, L, LAE); upper S slopes Lake Loloru crater, *Craven & Schodde 207* ♀ (A, BISH, BRI, CANB, L, LAE); vicinity of Kugugai, *Schodde & Craven 3980* ♂ (A, BRI, CANB, K, L, LAE); Koniguru, Buin, *Kajewski 2009* ♀ (A, BISH, BM, BO, BRI, G, L, P, SING).—FAURO: Barahono River area, *BSIP 13864* ♂ (L, LAE, SING); Haliuna River area, *BSIP 13963* ♂ & ♀ (L, LAE, SING).—SHORTLAND: Kupala River area, *BSIP 13163* ♂ (L, LAE, SING); northern Shortland Island, *BSIP 5862* ♂ (L, LAE, SING); SE Koangkopi River, *BSIP 13104* ♀ (L, LAE, SING).—TREASURY: Maloaini area, *BSIP 14290* ♂ (L, LAE, SING).—CHOISEUL: Pemba area, *BSIP 18754* ♂ (L, LAE, SING); E Sekoloe River, *BSIP 18604* ♀ (L, LAE, SING); Ologhata area, *BSIP 17525* ♀ (L, LAE, SING); Roke River, *BSIP 5393* ♀ (L, LAE, SING); Oaka River area, *BSIP 18423* ♂ & ♀ (L, LAE, SING); opposite Bembalama Island, *BSIP 4032* ♀ (L, LAE, SING).—ROB ROY: without further locality, *BSIP 5367* ♀ (L, LAE, SING).—BARORA FA: Pazeghere area, *BSIP 16249* ♂ (L, LAE, SING); Mt. Sears area, *BSIP 16528* ♀ (L, LAE, SING); Isaisao Point area, *BSIP 16609* ♂ (L, LAE, SING).—BARORA ITE: without further locality, *BSIP 16010* ♂ (L, LAE, SING).—SANTA YSABEL: Binusa, *BSIP 6770* ♂ (L, LAE, SING); Kolokofa River, *BSIP 5103* ♀ (L, SING); Kakatio, *Brass 3244* ♂ (A, BISH, BM, BO, BRI, G, L, SING); Allardyce Harbour, *BSIP 8275* ♂ (L, LAE, SING); near Tiratona, *BSIP 2285* ♀ (L, LAE, SING); Pilena Bay, *BSIP 6678* ♀ (L, LAE, SING).—MALAITA: Alasa Mtn, *Roy. Soc. Exped. 221* ♀ (A, K, L, LAE); Horamou, *BSIP 13670* ♀ (L, LAE, SING).—VELLA LAVELLA: Oula River area, *BSIP 11193* ♀ (L, LAE, SING).—RANONGGA: SW Ranongga, *BSIP 15627* ♀ (L, LAE, SING); Dae area, *BSIP 14443* ♀ (L, LAE, SING).—NEW GEORGIA: Roviana Lagoon, Baharo River, *BSIP 2021* ♂ (K, L, LAE, SING); Mango River, *BSIP 7123* ♀ (L, LAE, SING); Tita River, *BSIP 6006* ♀ (L, LAE, SING); Viru River, *BSIP 7150* ♂ (L, LAE, SING); Kalena Bay, *BSIP 5067* ♀ (K, L, LAE, SING).—VANGUNU: Balavaeni, *BSIP 6121* ♀ (L, LAE, SING); near mouth of Gevala River, *BSIP 907* ♀ (L, LAE, SING); Jalire, *BSIP 6424* ♂ (L, LAE, SING).—RENDOVA: W coast, 1 mi inland from Kenelo Plantation, *BSIP 1884* ♀ (L, LAE, SING).—TETEPARI: Evara River area, *BSIP 15888* ♀ (L, LAE, SING); Memesengo Point, *BSIP 15982* ♂ & ♀ (L, LAE, SING), *BSIP 15952* ♀ (L, LAE, SING).—GUADALCANAL: Vulolo, Tutuve Mtn, *Kajewski 2541* ♀ (A, BISH, BM, BO, BRI, G, P); Kambi Hill, Duidui, *BSIP 10107* ♂ (L, LAE, SING); Wanderer Bay area, *BSIP 12323* ♀ (L, LAE, SING).

The bark is reported as smooth to pustular or fissured, brownish to grayish; the wood is white to pale brown or yellowish. A preparation of macerated bark is rubbed on the skin to relieve stomach pains (*Kajewski 2541*).

14b. Timonius longitubus var. **pubescens** S. Darwin, var. nov.—TYPE: PAPUA NEW GUINEA. [Bismarck Archipelago.] New Britain: Turlu River (Gasmata

Subdistr.), in poor secondary forest on limestone, 1060 m, 25 Mar 1965, *NGF 24225* ♀ (holotype: A!; isotypes: BISH! BRI! CANB! LAE! SING!).

Idem ac varietas longitubus sed foliorum laminis sine domatiis et utrinque pubescentibus, stipulis abaxialiter effuse pubescentibus, et pedunculis post anthesin 3–4 mm latis differt.

Stipules abaxially pubescent with dense, spreading hairs. Leaf blades persistently pubescent above with short, spreading hairs scattered over lamina, densely pubescent beneath with spreading hairs; domatia absent. Peduncles of pistillate inflorescences usually 3–4 mm wide toward middle and densely pubescent. Flowers and mature fruits not seen.

Timonius longitubus var. *pubescens* is thus far known only from the type collection (locality shown in Fig. 16), taken from a leafy, much-branched tree 30 ft tall (bole 15 ft) and 6 inches dbh with open crown; the collection bears immature fruits. In addition to characters mentioned above in the key, the leaf venation of var. *pubescens* is somewhat less conspicuously paxillate than that of var. *longitubus*.

15. Timonius meridionalis S. Darwin, sp. nov.—TYPE: PAPUA NEW GUINEA. Milne Bay: Goodenough Island, east slopes, 1600 m, 16 Oct 1953, *Brass 24773* ♀ (holotype: A!; isotypes: CANB! K! L! LAE!).

Timonius a speciebus ad subgenus Abbottii ascriptis combinatione stipularum cinereopuberularum cum foliorum laminis ellipticis glabris vel sparse puberulis sine domatiis marsupiformibus et cum pyrenis oblique dispositis differt.

Epiphytic shrubs to free-standing trees to 30 m tall and 30 cm dbh; nature of bark and wood unknown. Branchlets 2–3 mm wide toward apex, pale to dark gray, glabrous to minutely puberulent. Stipules to 40 mm long and 10 mm wide, lanceolate, acuminate, minutely gray-puberulent to glabrous abaxially, glabrous adaxially except for colleters near base. Petioles 5–20 (–25) mm long, 1–2 mm wide at middle, semi- to subterete, glabrous; leaf blades 3–12 cm long, 1–6 cm wide, elliptic, sharply acuminate at apex, acute to attenuate at base, stiff-coriaceous, glabrous above, glabrous or with scattered appressed hairs over lamina beneath, the costa plane above, elevated beneath, the secondary nerves indiscernible, the veinlets prominulous above and beneath, paxillate; domatia cryptlike and porate, distributed along costa and also scattered over lamina beneath, glabrous to pubescent around ostiole. Staminate inflorescences 5- to 7-flowered, at anthesis to 2 cm long and 2 cm wide; peduncles 2–8 (–10) mm long, ca. 1.5 mm wide, compressed, glabrous; bracts absent. Staminate flowers: calyx limb cupular, truncate to shallowly undulate, glabrous or ciliate, the tube 2–2.5 mm long from base; corolla narrowly infundibular, white, the tube to 11 mm long, 1–1.5 mm wide, glabrous outside, glabrous within, the lobes 4, ca. 2.5 mm long, 1.5 mm wide, ovate, acute, papillose, the corolla limb 5–8 mm in diameter; stamens 4, the filaments short, the anthers ca. 5 mm long, 0.5 mm wide, linear, glabrous; pistillodium ca. 4 mm long, glabrous. Pistillate inflorescences 1- (2-)flowered; peduncles 2–5 (in fruit to 8) mm long, ca. 1 mm wide, compressed, glabrous; bracts absent. Pistillate flowers: calyx limb cupular, truncate, glabrous, the tube 0.5–1 mm long; corolla narrowly to broadly infundibular-cylindric, white, the tube 6–8 mm long, 1.5–2 mm wide,

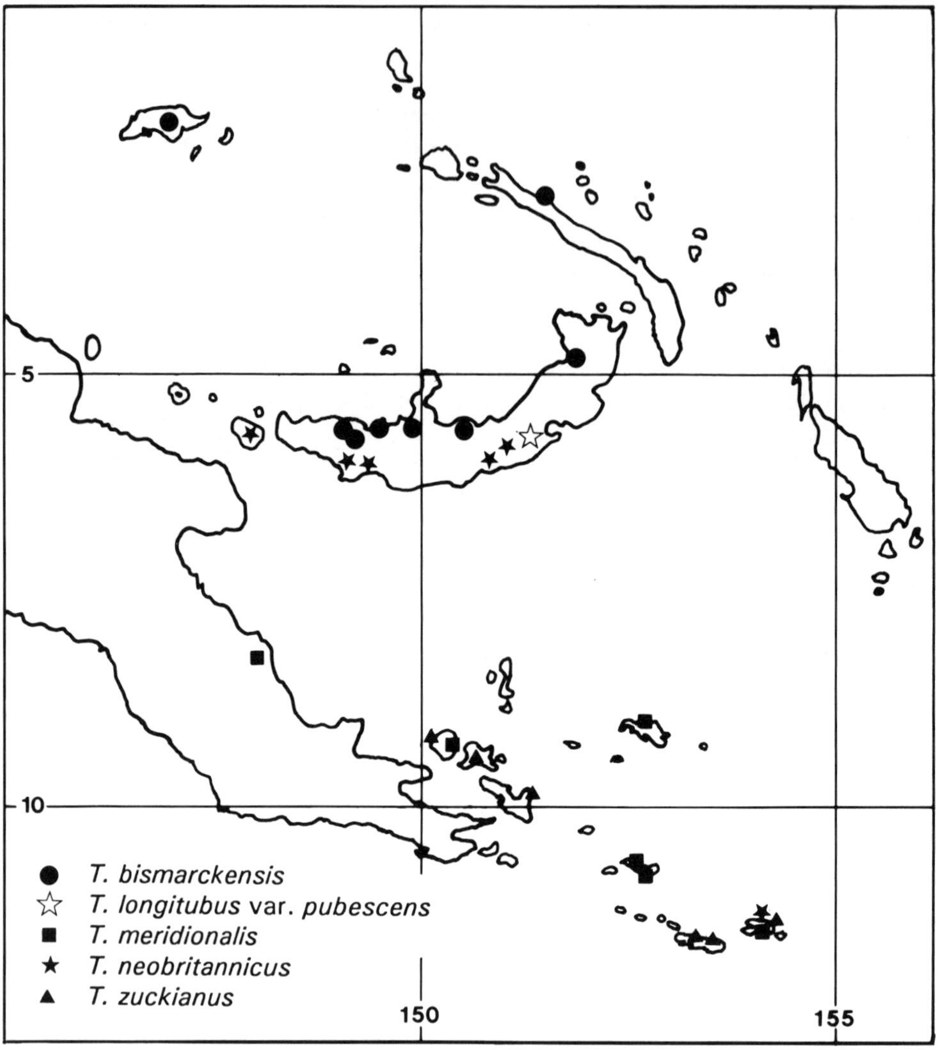

FIG. 16. Distribution of *Timonius bismarckensis*, *T. longitubus* var. *pubescens*, *T. meridionalis*, *T. neobritannicus*, and *T. zuckianus*.

glabrous outside, glabrous within, the lobes 7–8, 2–3 mm long, ca. 1.5 mm wide, lanceolate, acute, papillose, the corolla limb 6–8 mm in diameter; staminodia 7–8, ca. 2.5 mm long, linear, glabrous; hypanthium 2–3 mm long, 2–3.5 mm in diameter, subglobose, glabrous, the disc glabrous, the style 8–10 mm long, 7- to 8-armed, glabrous. Fruits 4–7 mm long, 6–10 mm in diameter, depressed-globose, tuberculate toward base when dry, glabrous, color of mature fruits unknown; pyrenes 40–70 in 8 radiating double-files, subequal, obliquely oriented. Figs. 3n, 17.

Phenology. Flowering and fruiting throughout the year.

Distribution (Fig. 16). Southeastern New Guinea and Papuan Islands; moist primary or sometimes secondary forests; 60–1600 m.

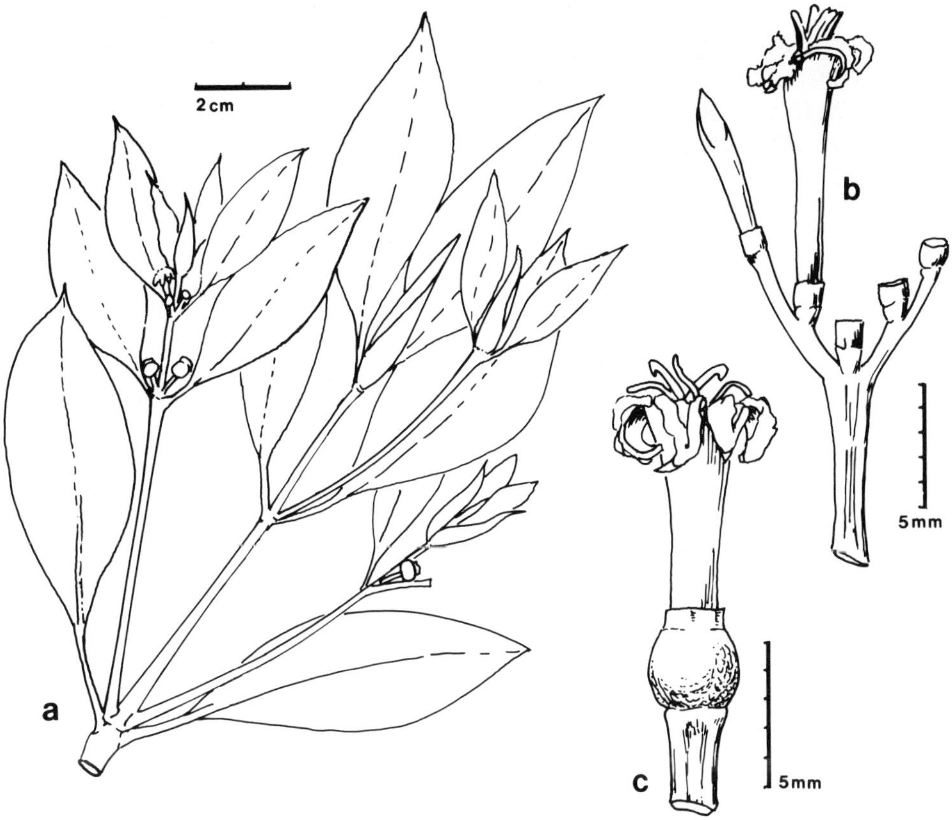

FIG. 17. *Timonius meridionalis*. a. Branchlets with pistillate flowers and fruits. b. Staminate inflorescence. c. Pistillate flower. (Based on: a, c, *Brass 27532*; b, *Brass 28630*.)

ADDITIONAL SPECIMENS EXAMINED. **Papua New Guinea.** [*Papua.*] NORTHERN: N of Ioma on track to Nindewari, *NGF 29977* ♀ (A, CANB, E, K, L, LAE, SING).—MILNE BAY: Goodenough Island, E slopes, *Brass 24796* ♂ (A, CANB, L, LAE), *Brass 25082* ♀ (A, CANB, L, LAE); Woodlark Island, Kulumadau, *Brass 28630* ♂ (A, K, L, LAE); Misima Island, Mt. Sisa, N slopes, *Brass 27460* ♀ (A, CANB, K, L, LAE), *Brass 27532* ♀ (A, CANB, K, L, LAE); Misima Island, foot of Mt. Oeatau, *LAE 74603* ♀ (A, CANB, E, L, LAE, QRS); Rossel Island, Mt. Rossel, S slopes, *Brass 28449* ♀ (A, K, L, LAE).

The delimitation of *T. meridionalis* is not entirely satisfactory, and while the collections assigned here are very similar in vegetative characters, especially in having gray-pubescent stipules and elliptic leaf blades lacking large domatia, they differ in stipule size, shape of pistillate corolla, and size of fruit (many of which may be immature); it is entirely possible that more than one species is represented. *Timonius meridionalis* is most similar to *T. pseudaffinis* and *T. neobritannicus*, both of which are also distributed over land areas adjacent to the Solomon Sea. *Timonius pseudaffinis* differs in having smaller stipules (to 25 mm long), and more slender petioles (usually less than 1 mm wide) and peduncles (less than 1 mm wide). *Timonius neobritannicus* has generally smaller leaves (blades to 3 cm wide), smaller fruits (ca. 4 mm long and 6 mm in diameter), and pistillate flowers with 6 lobes. A specimen from the Mt. Victoria area of Papua, *van Royen 10985* ♂ (BISH, BRI, LAE), collected at 2670 m, approaches *T. meridionalis*, but may be more closely allied with *T. wollastonii*, another species of high elevations but with thick-coriaceous and retuse leaf blades (cf. van Royen, Alpine Fl. New Guinea 4: 2780, *pl. 190*. 1983).

16. Timonius modestus Merrill & Perry, J. Arnold Arb. 26: 237, fig. 2B. 1945.—TYPE: [PAPUA NEW GUINEA.] Papua. Western: Fly River, 528 mile camp, 80 m, May 1936, *Brass 6748* ♀ (holotype: A!; isotypes: BM! BO! BRI! L!).

Epiphytic shrubs to small trees, rarely to 20 m tall; nature of bark and wood unknown. Branchlets 2–3 mm wide toward apex, dark red-brown, minutely puberulent or canescent to glabrous. Stipules to 70 mm long and 12 mm wide, long-lanceolate, acuminate, canescent with whitish hairs to glabrous abaxially, glabrous to subciliate adaxially except for colleters near base. Petioles 5–15 mm long, to 2 mm wide at middle, semiterete, canescent to glabrous; leaf blades 4–13 cm long, 2–6 cm wide, obovate to elliptic, bluntly to sharply acute to acuminate at apex, attenuate at base, coriaceous, glabrous above, glabrous beneath, the costa plane above, elevated beneath, the secondary nerves hardly discernible, the veinlets prominulous above, paxillate; domatia small open pits scattered over lamina and often also pocket- or cryptlike along margin of lamina beneath, glabrous to puberulent. Staminate inflorescences 5- to 7-flowered, at anthesis to 2 cm long and 3.5 cm wide; peduncles 5–10 mm long, to 1.5 mm wide, compressed, canescent to subglabrous; bracts absent. Staminate flowers: calyx limb cupular, truncate, glabrous, the tube 2–2.5 mm long from base; corolla infundibular, white to greenish, the tube 9–10 mm long, 1–1.5 mm wide, grayish puberulent outside, glabrous within, the lobes 4, 2–3 mm long, 1–1.5 mm wide, ovate, acute, papillose, the corolla limb to 7 mm in diameter; stamens 4, the filaments short, the anthers 5–6 mm long, linear, glabrous; pistillodium not seen. Pistillate inflorescences 1-flowered; peduncles 5–12 mm long, 1–2 mm wide, compressed, canescent to glabrous; bracts absent. Pistillate flowers: calyx limb cylindric, truncate, glabrous, the tube 0.5–1.5 mm long; corolla infundibular, white to greenish, the tube (late bud) to 5 mm long, ca. 2.5 mm wide, puberulent to pale-strigillose outside, glabrous within, the lobes 8–10, 3–4 mm long in bud, subulate, acute, the corolla limb 6–10 mm in diameter; staminodia 6–10, 2.5–3 mm long, ca. 0.2 mm wide, linear, glabrous; hypanthium 2.5–3 mm long, 3.5–4.5 mm in diameter, subglobose, glabrous, the style irregularly 8- to 10-armed, glabrous. Fruits 5–6 mm long, 6.5–9 mm in diameter, depressed-globose, somewhat laterally compressed, tuberculate, glabrous; pyrenes 70–120 in as many as 12 radiating double-files, subequal, obliquely oriented.

Phenology. Collected in flower intermittently from June to October, in fruit from February to September.

Distribution (Fig. 9). Widely distributed in New Guinea; in primary, often wet forest; 40–600 m.

ADDITIONAL SPECIMENS EXAMINED. **Indonesia.** IRIAN JAYA [*Western New Guinea*]: Geelvink Bay, Nabire, *Kanehira & Hatusima 11667* ♀ (A, L); Dalman, 45 km inward of Nabire, *Kanehira & Hatusima*

12268 ♀ (A, BO).—MIMIKA: Bivakeiland, *Pulle 42* ♀ (BO, K, L), *Pulle 456a* (A, L). **Papua New Guinea.** [*Papua*.] SOUTHERN HIGHLANDS: near Waro airstrip, 20 km SSW of Kutubu, *Jacobs 9230* ♂ (BISH, L, LAE).—WESTERN: Palmer River, 2 mi below junction with Black River, *Brass 7076* ♂ (A, BO, BRI, G, K, L); Rumaganae Road, 5 mi from Kiunga, *NGF 43844* ♀ (A, BISH, BO, CANB, K, L, LAE, SING).

The circumscription of *T. modestus* is as yet vague; the type and other collections from south-central New Guinea have relatively large leaf blades lacking the prominent pocketlike domatia that flank the abaxial costa in, for example, *T. belensis*. In the pale stipule pubescence and domatia characters, *T. modestus* resembles *T. meridionalis* and *T. bismarckensis*; two staminate specimens from northeastern New Guinea [Madang Province, Gati-Berg, 800 m, *Schlechter 16860* (A, BRI, C, G, L, S); Morobe Province, Jaduna, 200 m, *Schlechter 19304* (BM, C, G, K, S)] approach *T. modestus*, but may be referable to one of the two last-named species. A collection from the Vogelkop Peninsula (*BW 12377* ♂) and two from Morotai (*Kostermans 1098* ♀ and *1107* ♀) resemble *T. modestus*, but have more robust branchlets and larger fruits; they are here tentatively referred to *T. modestus*.

17. Timonius neobritannicus S. Darwin, sp. nov.—TYPE: PAPUA NEW GUINEA. Morobe: Umboi Island, Liplip–Mambi River area, epiphyte 20 m tall in lowland rain forest, 200 m, 5 Oct 1974, *LAE 66113* ♀ (holotype: A!; isotypes: E! K! L! LAE! M!).

Timonius ex affinitate T. pseudaffinis S. Darwin sed domatiis saepe magnis marsupiformibus utrinque basin costae versus et stipulis majoribus et pedunculis latioribus differt.

Epiphytic shrubs becoming trees to 15 cm dbh. Bark gray to brown and scaly, the underbark red-brown, the wood white and hard. Branchlets 1–2 mm wide toward apex, pale, glabrous. Stipules to 40 mm long and 8 mm wide, lanceolate, acuminate, glabrous to ciliate to gray appressed-puberulent abaxially, glabrous adaxially except for colleters near base. Petioles 3–20 mm long, 1–2 mm wide at middle, semi- to subterete, glabrous to scattered-puberulent; leaf blades 4–9 cm long, 1–3 cm wide, oblanceolate to narrowly elliptic, sharply acute to acuminate at apex, acute to attenuate at base, firm-coriaceous, glabrous above, glabrous beneath, the costa plane to canaliculate above, elevated beneath, the secondary nerves indistinguishable, the veinlets prominulous above, paxillate; domatia pocketlike or cryptlike, large, solitary or usually a pair at base of blade, or these also along costa or rarely distributed over lamina beneath, rarely absent, glabrous or with few marginal hairs. Staminate inflorescences 5-flowered; peduncles to 8 mm long, ca. 1.5 mm wide, compressed, glabrous; bracts absent. Staminate flowers: calyx limb cupular-cylindric, truncate to undulate, glabrous or with short, spreading hairs at apex of abortive ovary, the tube to 2 mm long from base; corolla narrowly infundibular, the tube 7–9 mm long, ca. 1.5 mm wide, glabrous outside, glabrous within, the lobes 4, 3–4 mm long, ca. 1 mm wide, ovate, acute, somewhat papillose, the corolla limb 7–10 mm in diameter; stamens 4, the filaments short, the anthers ca. 6 mm long, 0.25 mm wide, linear, glabrous; pistillodium not seen. Pistillate inflorescences 1-flowered; peduncles 4–6 mm long, ca. 1.5 mm wide, compressed, glabrous; bracts absent. Pistillate flowers: calyx limb cupular, truncate, glabrous, the tube ca. 0.5 mm long; corolla cylindric-infundibular, the tube ca. 4 mm long, ca. 1.5 mm wide, glabrous outside, glabrous within, the lobes 6, ca. 2.5 mm long, 1 mm wide, ovate-lanceolate, acute, somewhat papillose, the corolla limb ca. 5 mm in diameter; staminodia not seen; hypanthium 2–3 mm long, 2.5–3.5 mm in diameter, depressed-globose, tuberculate-costate, glabrous, the style 6-armed. Fruits ca. 4 mm long, ca. 6 mm in diameter, depressed-globose, glabrous; pyrenes 40–50 in 6–8 radiating double-files, subequal, obliquely arranged. Figs. 3j, 4a, b, 18.

Phenology. Probably flowering and fruiting throughout the year.

Distribution (Fig. 16). Bismarck Archipelago (south coast of New Britain, Umboi Island) and Rossel Island; in lowland primary forests to ridgetop forests; 0–2000 m.

ADDITIONAL SPECIMENS EXAMINED. **Papua New Guinea.** [*Bismarck Archipelago.*] NEW BRITAIN: 5 mi N old site of Omoi, Arawe area, *NGF 26517* ♂ (LAE); Pulie River, W side, *NGF 27241* ♂ (A, CANB, K, L, LAE, SING); Pirilongi, *NGF 21945* ♂ (A, CANB, K, L, LAE, SING); Mt. Lululua, 30 mi NE Fulleborn Harbour, *NGF 34411* ♂ (A, K, L, LAE, M).—[*Papua.*] MILNE BAY: Rossel Island, Mt. Rossel, *LAE 76014* ♀ (A, K, L, LAE, M, QRS).

Timonius neobritannicus resembles *T. meridionalis* and *T. pseudaffinis*, all in adjacent land areas surrounding the Solomon Sea. *Timonius pseudaffinis* resembles *T. neobritannicus* in its usually oblanceolate leaf blades with prominulous venation, as well as its relatively small fruits, but differs in having pitlike domatia (not a pair of large, pocketlike domatia flanking the abaxial costa), stipules to ca. 25 mm long, and petioles less than 1 mm wide. In *T. meridionalis*, the leaf blades are elliptic, rather than oblanceolate, among other differentiating characters mentioned in the key. *Timonius neobritannicus* is noted on specimen labels as growing in association with *Calophyllum*, *Metroxylon*, *Garcinia*, and *Casuarina*.

18. Timonius oktediensis S. Darwin, sp. nov.—TYPE: PAPUA NEW GUINEA. Western: Ok Tedi headwaters near Kennecott field camp, 853 m, 29 Oct 1969, *NGF 42793* ♂ (holotype: LAE!; isotypes: A! CANB! K! L! SING!).

Timonius a speciebus subgeneris Abbottiae combinatione corollarum pubescentium cum foliis basi rotundatis et calycibus masculis ciliatis distinguendus.

Woody epiphytes; nature of bark and wood unknown. Branchlets 3–5 mm wide toward apex, brown, glabrous. Stipules to 60 mm long and 12 mm wide, lanceolate, acuminate, glabrous abaxially, puberulent with appressed hairs adaxially toward middle and base. Petioles to 4 mm long, ca. 2 mm wide at middle, semi- to subterete, glabrous; leaf blades 3–9 cm long, 1–4 cm wide, elliptic, acute to abruptly and sharply acuminate at apex, tapering and abruptly rounded at base, stiff-coriaceous, glabrous above, glabrous and rather lustrous beneath, the costa subplane above, elevated beneath, the secondary nerves indistinct, the veinlets submerged to prominulous above and beneath, paxillate; domatia small, pitlike, open or angled, distributed over lamina beneath, larger and pocketlike along costa, or sometimes absent, glabrous. Staminate inflorescences 3- to 5-flowered, at anthesis to 2.5 cm long and 2 cm wide; peduncles 6–12 mm long, ca. 2 mm wide, strongly compressed, glabrous or with scattered, short, spreading hairs; bracts absent. Staminate flowers: calyx limb cupular, truncate, ciliate but otherwise

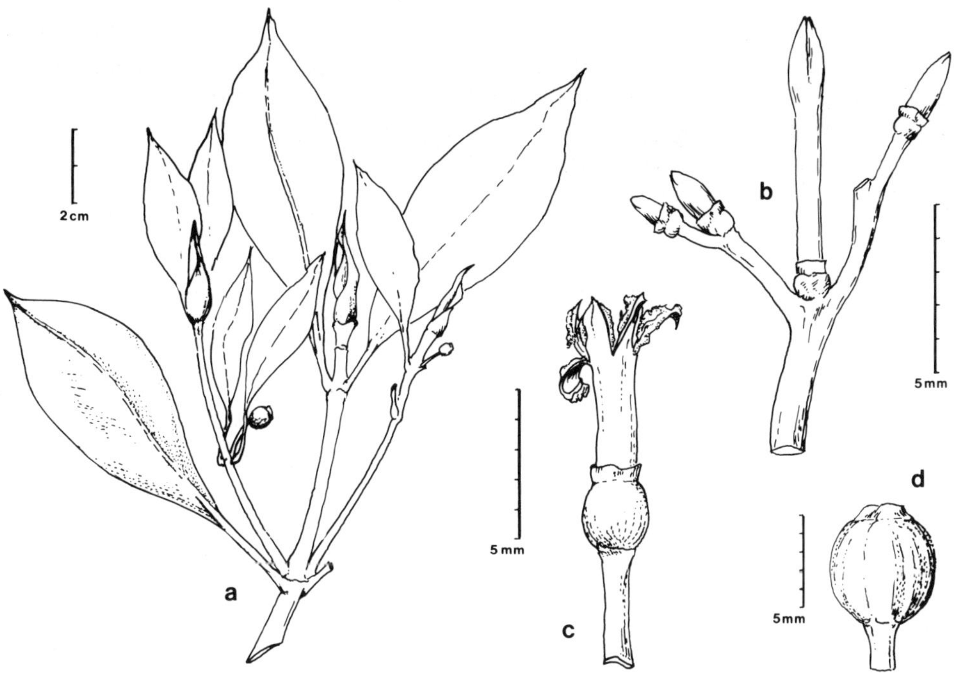

FIG. 18. *Timonius neobritannicus*. a. Branchlets with fruits. b. Staminate inflorescence. c. Pistillate flower. d. Fruit. (Based on: a, d, *LAE 76014*; b, *NGF 34411*; c, *LAE 66113*.)

glabrous, the tube to 2.5 mm long from base; corolla hypocrateriform, the tube to 10 mm long, 1.5–2 mm wide, sparsely spreading-puberulent outside, glabrous within, the lobes 4, ca. 4 mm long, 1–1.5 mm wide, lanceolate, acute, papillose; stamens 4, the filaments short, the anthers 6–7 mm long, ca. 0.5 mm wide, linear, glabrous; pistillodium ca. 4 mm long, linear, terete, glabrous. Pistillate inflorescences not seen. Fruits not seen. Figs. 3m, 15e, f.

As yet known only from a single staminate collection (locality shown in Fig. 19), *T. oktediensis* is nonetheless distinctive in having rounded leaf blade bases, ciliate calyces, and corollas that are sparsely spreading-pubescent. In the absence of mature fruits, *T. oktediensis* cannot be positively aligned with other species of subg. *Abbottia*, although the adaxial stipule pubescence might suggest *T. pubistipulus*, which also occurs in central New Guinea.

19. Timonius pachyphyllus Merrill, Philipp. J. Sci. 26: 494. 1925.—TYPE: PHILIPPINES. Luzon: Rizal Prov., Mount Angilog, 20 Apr 1922, *Philipp. Bur. Sci. 40794* ♀ (holotype: PNH, destroyed; isotype: A!).

Shrubs ca. 4 m tall; nature of bark and wood unknown. Branchlets 4–6 mm wide toward apex, dark becoming pale brown, glabrous or with scattered, lax hairs, these more abundant at nodes above stipule scars. Stipules to 45 mm long and 15 mm wide, caducous but the very base persisting as a low collar around node, ovate, acute to obtuse, glabrous or with sparsely scattered hairs abaxially, glabrous adaxially except densely pubescent in axil and with colleters along midrib and near

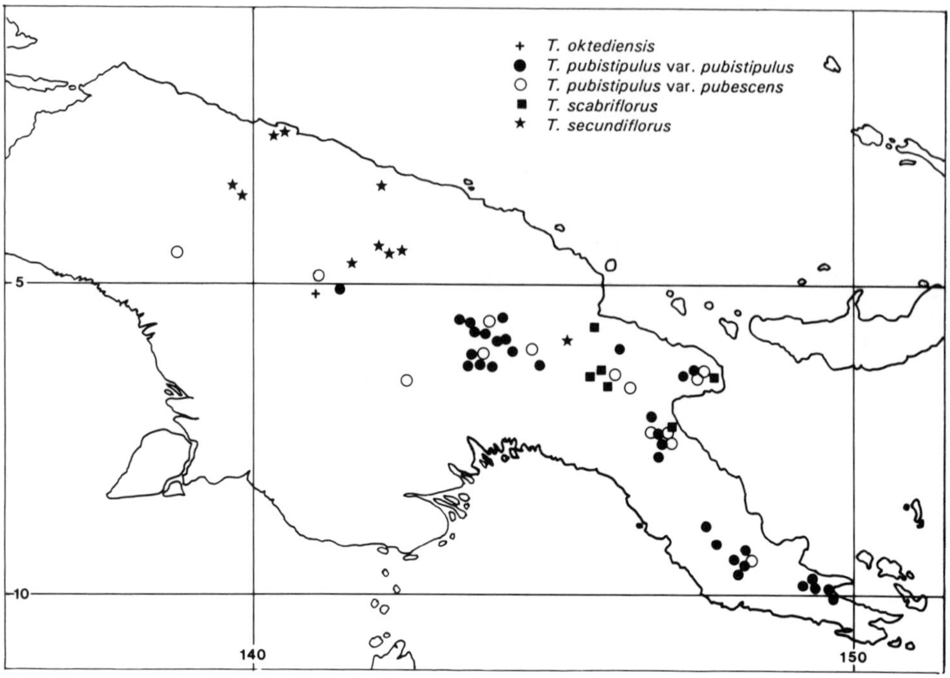

FIG. 19. Distribution of *Timonius oktediensis*, *T. pubistipulus* var. *pubistipulus*, *T. pubistipulus* var. *pubescens*, *T. scabriflorus*, and *T. secundiflorus*.

base. Petioles 10–20 mm long, 3–4 mm wide at middle, winged owing to the decurrent leaf blade, glabrous or with scattered, lax hairs; leaf blades 8–14 cm long, 3–5 cm wide, elliptic to subfalcate, narrowed toward the apex but rounded at the tip, long-attenuate at base, coriaceous, glabrous or with scattered, lax hairs above and beneath, the costa plane to canaliculate above, elevated beneath, the secondary nerves hardly distinguishable, ca. 9 on each side of costa, the veinlets submerged beneath, striate above, paxillate; domatia cryptlike and porate, distributed on lamina beneath, glabrous, often absent. Staminate inflorescences (5-) 7- to 9-flowered, at anthesis to 4 cm long and 4 cm wide; peduncles 12–20 mm long, 2–3 mm wide, compressed to strap-shaped, subglabrous; bracts absent. Staminate flowers: calyx limb cupular, truncate to undulate, subglabrous, the tube 2.5–3.5 mm long from base; corolla narrowly infundibular, the tube 9–11 mm long, 1.5–2 mm wide, glabrous outside, glabrous within, the lobes 4, 4–4.5 mm long, 2–2.5 mm wide, ovate-elliptic, bluntly acute and with 3 irregular, fingerlike appendages to 1.2 mm long extending from subapical interior margin, the corolla limb to 6 mm in diameter; stamens 4, the filaments short, the anthers ca. 6 mm long, 0.6 mm wide, linear-oblong to -clavate, glabrous; pistillodium to 5 mm long, ca. 0.2 mm in diameter, filiform, glabrous. Pistillate inflorescences 2- to 3-flowered, at anthesis to 3 cm long and 4 cm wide; peduncles 10–15 mm long, ca. 1.5 mm wide, compressed, glabrous or with scattered hairs; bracts absent. Pistillate flowers: calyx limb cupular, truncate to undulate, subglabrous, the tube 1–2 mm long; corolla infundibular, the tube 6–8 mm long, ca. 1.5 mm wide, glabrous outside, glabrous within, the lobes 6, 2.5–3 mm long, 1.5–2 mm wide, ovate-elliptic, bluntly acute with appendages as in corolla of staminate flowers, the corolla limb to 6 mm in diameter; staminodia not seen; hypanthium 2.5–4 mm long, 2.5–3 mm in diameter, ellipsoid, subglabrous, the disc glabrous, the style not seen. Fruits 5–6 mm long, 6–7 mm in diameter, subglobose to laterally compressed, glabrous, color of mature fruits unknown; pyrenes ca. 40 in 6 radiating double-files, subequal, obliquely oriented. Figs. 3d, 15a.

Distribution. Philippines, Luzon, Rizal Province; forested slopes; ca. 1200 m.

Additional Specimen Examined. **Philippines.** Luzon: Rizal Prov., Mt. Irig, *Philipp. Bur. Sci. 42277* ♂ (A).

Timonius pachyphyllus is closely related to *T. appendiculatus*. Both are the only species of subg. *Abbottia* known from the Philippine Islands; both have pistillate flowers arranged in cymose inflorescences; and both have corolla lobes invested with subterminal, adaxial appendages. *Timonius pachyphyllus* differs in its larger, thicker, more obtuse leaf blades and the fewer-flowered inflorescences, among other characters.

20. Timonius pseudaffinis S. Darwin, sp. nov.—Type: Solomon Islands. Santa Ysabel: Tatamba, ca. 10 m, climber on old garden land, fruits deep green, 5 Oct 1965, *Royal Society Exped. (P. F. Hunt) 2876* ♀ (holotype: A!; isotypes: K! L! LAE!).

Timonius ex affinitate T. affinis A. Gray sed ab ramulis petiolis pedunculisque gracilioribus et foliorum laminis plerumque oblanceolatis differt.

Epiphytic shrubs, "climbers," or "stranglers," or trees to 40 m tall; nature of bark and wood unknown. Branchlets 1–2 mm wide toward apex, dark gray to reddish brown, glabrous. Stipules to 25 mm long and 7 mm wide, lanceolate, acute, glabrous to subciliate abaxially, glabrous adaxially except for colleters near base. Petioles 2–15 mm long, 0.5–1 mm wide at middle, semi- to subterete, glabrous; leaf blades 3–11 cm long, 1–3.5 cm wide, oblanceolate, acute to acuminate at apex, acute at base, chartaceous to subcoriaceous, glabrous above, glabrous beneath, the costa plane to canaliculate above, prominulous beneath, the secondary nerves hardly discernible, 6–8 on each side of costa, spreading, the veinlets prominulous, paxillate; domatia pitlike, distributed along costa or absent, glabrous to puberulent. Staminate inflorescences 3- to 5- (11-)flowered, at anthesis to 2 cm long and 2 cm wide; peduncles 2–7 mm long, ca. 0.5 mm wide, glabrous; bracts absent. Staminate flowers: calyx limb cupular, truncate to undulate, glabrous, the tube 1–1.5 mm long from base; corolla narrowly infundibular to hypocrateriform, white to yellowish or reddish, the tube to 11 mm long, 1–1.2 mm wide, glabrous outside, glabrous within, the lobes 4, 2–2.5 mm long, 1–1.5 mm wide, lanceolate, acute, essentially glabrous, the corolla limb to 6 mm in diameter; stamens 4, the filaments to 0.2 mm long, the anthers ca. 5 mm long, 0.5 mm wide, linear-sagittate, glabrous; pistillodium to 2 mm long, filiform, glabrous. Pistillate inflorescences 1-flowered; peduncles to 7 mm long, to 1 mm wide, glabrous; bracts absent or hardly discernible. Pistillate flowers: calyx limb cupular, truncate, glabrous, the tube to 1 mm long; pistillate corolla not seen; hypanthium to 3 mm long, ca. 2 mm in diameter, ellipsoid to subglobose, glabrous, the disc glabrous, the style to 11 mm long, ca. 0.8 mm wide near middle,

4-armed, glabrous. Fruits 5–9 mm long, 5–9 mm in diameter, subglobose, tuberculate, glabrous, color of mature fruits unknown; pyrenes 18–25 in 4 radiating double-files, subequal, obliquely oriented. Figs. 3k, 6j, 20a–c.

Phenology. Flowering and fruiting throughout the year.

Distribution (Fig. 12). Solomon Islands, from Rob Roy Island southward to San Cristobal; often common in primary or secondary forests; 10–1100 m.

Vernacular names. *Simidi, Semende, Sement, Aisemende, Aisimidi,* or *Haisirihan* (Kwara'ae).

ADDITIONAL SPECIMENS EXAMINED. **Solomon Islands.** ROB ROY: *BSIP 5349* ♂ (K, L, LAE, SING).—KOLOMBANGARA: Rei area, *BSIP 11654* ♀ (L, LAE, SING); Bambari area, *BSIP 7519* ♀ (L, LAE, SING); Shoulder Hill area, *BSIP 11474* ♂ (K, L, LAE, SING); Sandy Harbour, *BSIP 1498* ♀ (L, LAE); W of Vila River, *BSIP 8526* ♀ (L, LAE, SING).—NEW GEORGIA: Tita River, Viru Harbour, *BSIP 2536* ♂ (K, L, LAE); Viru Harbour, *BSIP 5192* ♀ (L, LAE, SING).—RENDOVA: first main ridge N of Kenelo Plantation, *BSIP 1902* ♂ (L, LAE, SING).—VANGUNU: SE coast between Vura and Gevala River, *BSIP 869* ♂ (L, LAE, SING).—SANTA YSABEL: Maringe Lagoon, summit Mt. Sasari, *BSIP 2387* ♂ (L, LAE); Tatamba, *BSIP 2690* ♂ (L, LAE, SING).—MALAITA: Are Are distr., Mota, 5 mi inland from Kiu, *BSIP 3907* ♀ (L, LAE, SING); N of Pululaha, *BSIP 16258* ♀ (K, L, LAE, SING).—BIG NGGELA: Hagela area, *BSIP 15191* ♀ (A, L, LAE, SING).—GUADALCANAL: Duidui area, *BSIP 12056* ♀ (L, LAE, SING); Makina area, Marau, *BSIP 9476* ♀ (L, LAE, SING).—SAN CRISTOBAL: headwaters of Huni River, 4 mi E of Wainoni, *Roy. Soc. Exped. 6319* ♂ (A, K, L, CANB, LAE).

Timonius pseudaffinis is readily distinguished from *T. longitubus*, which is distributed throughout the Solomon Islands but is much more robust in vegetative and reproductive parts. *Timonius pseudaffinis* is most similar to *T. affinis*, known from Vanuatu, the Santa Cruz Islands, Fiji, and Samoa, but *T. affinis* differs in having stouter branchlets (2–3 mm wide), petioles (1–2 mm wide at middle), and peduncles (8–25 mm long and 1–2 mm wide at anthesis). It also has elliptic, oblong, or obovate leaf blades (versus consistently oblanceolate blades in *T. pseudaffinis*) with the secondary nerves usually distinguishable from the veinlets.

21. Timonius pubistipulus S. Darwin, J. Arnold Arb. 64: 612. 1983.—TYPE: PAPUA NEW GUINEA. Western Highlands: Minj River valley above Nona River camp, in *Nothofagus* forest, 2750 m, 30 Jul 1957, *Pullen 268* ♀ (holotype: A!; isotypes: BISH! BM! BO! BRI! CANB! G! L! LAE! US!).

Epiphytic shrubs to lianas or terrestrial trees to 35 m tall and 50 cm dbh. In var. *pubistiulus*, bark gray and smooth to brownish and pustular or scaly, the wood reportedly white to reddish straw-colored; in var. *pubescens*, bark pale gray to dark brown and smooth, the wood reportedly white to pale brown. Branchlets 3–4 mm wide toward apex, glabrous to conspicuously pubescent. Stipules to 80 mm long and 15 mm wide, lanceolate, long-acuminate, glabrous abaxially, densely pale-sericeous adaxially. Petioles 2–20 mm long, 1.5–2 mm wide at middle, semi- to subterete, glabrous; leaf blades 2.5–12 cm long, 1–5.5 cm wide, elliptic to obovate, acute to short-acuminate at apex, acute to attenuate at base, stiff-coriaceous, glabrous above, glabrous beneath, the costa canaliculate above, elevated beneath, the secondary nerves indistinguishable, the veinlets prominulous above and beneath, paxillate; domatia small or large, pitlike, distributed over lamina beneath, glabrous to puberulent. Staminate inflorescences 3- to 5-flowered, at anthesis ca. 1.5 cm long

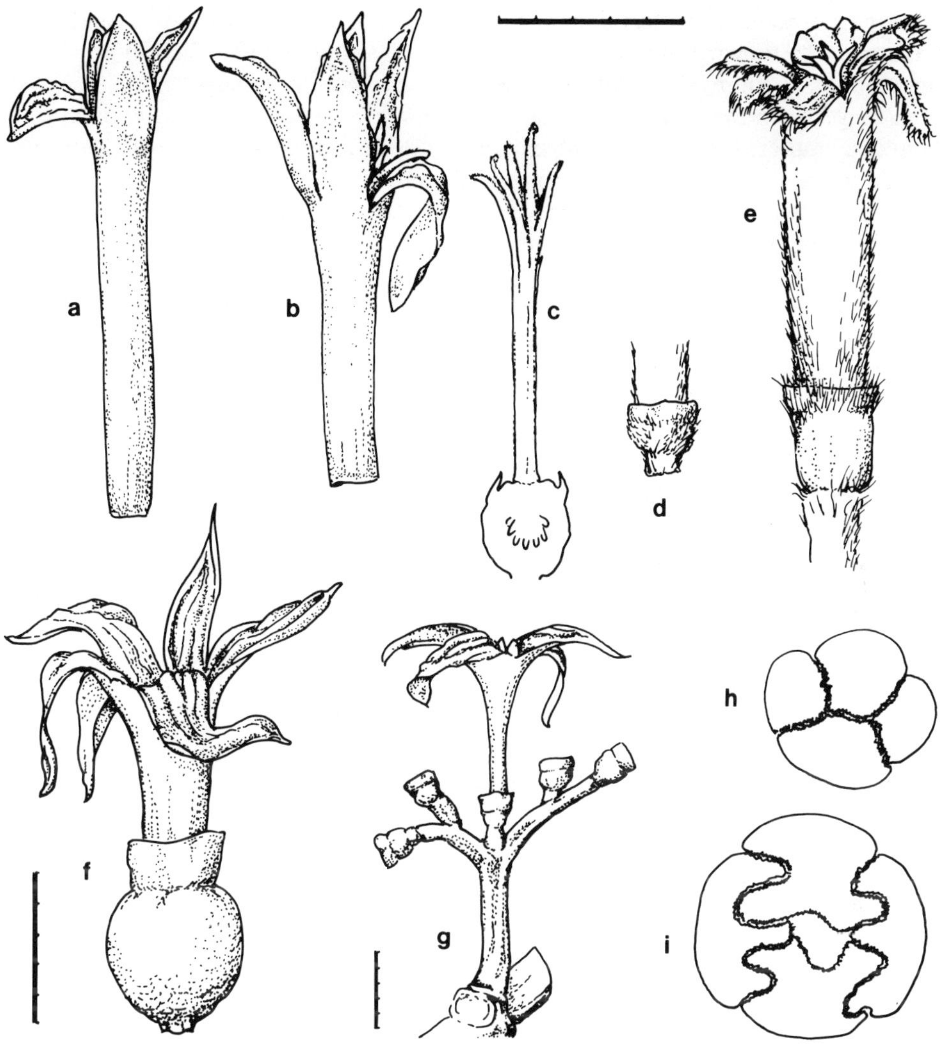

FIG. 20. *Timonius pseudaffinis*, *T. subavenis*, *T. trichanthus*, *T. zuckianus*, and *T. pubistipulus* var. *pubescens*. a–c. *T. pseudaffinis*. a. Corolla of staminate flower. b. Corolla of pistillate flower. c. Style and sectioned ovary of pistillate flower. d. *T. subavenis*, calyx of staminate flower. e. *T. trichanthus*, pistillate flower. f, g. *T. zuckianus*. f. Pistillate flower. g. Staminate inflorescence. h, i. *T. pubistipulus* var. *pubescens*. h. Cross section of corolla lobes of staminate flower near distal end of bud, magnified. i. Same, near middle of corolla lobes, showing interlocking aestivation. Bars = 5 mm. (Based on: a, *BSIP 2536*; b, c, *BSIP 1787*; d, *Pulle 650*; e, *Hartley 12533*; f, *Brass 28458*; g, *Brass 28389*; h, i, *LAE 77669*.)

and 0.8 cm wide; peduncles 3–5 mm long, 1.5–2 mm wide, compressed, glabrous. Staminate flowers: calyx limb cupular, truncate to undulate, glabrous, the tube 0.5–2 mm long from base; corolla infundibular, the tube to 8 mm long, ca. 1.5 mm wide, glabrous outside, glabrous within, the lobes 4, 2.5–3.5 mm long, ca. 1 mm wide, lanceolate, acuminate, glabrous to papillose, the corolla limb to 5 mm in diameter; stamens 4, ca. 6 mm long, glabrous; pistillodium not seen. Pistillate inflorescences 1-flowered; peduncles 1–5 (in fruit to 12) mm long, 1–2 mm wide, compressed, glabrous; bracts absent. Pistillate flowers: calyx limb cupular, truncate to undulate, glabrous, the tube 0.5–1 mm long; corolla broadly infundibular, the tube (3–) 4–7 mm long, 2–2.5 mm wide, glabrous outside, glabrous within, the lobes 6–8, 2.5–3.5 mm long, 1–1.5 mm wide, lanceolate, acuminate, papillose, the corolla limb to 8 mm in diameter; staminodia 6 (perhaps also 8?), ca. 2 mm long, sagittate, glabrous; hypanthium 3–3.5 mm long, 3–4 mm in diameter, subglobose, glabrous, the disc glabrous, the style irregularly 6- to 8-armed, glabrous. Fruits 5–9 mm long, 6–10 mm in diameter, subglobose to broadly obovoid and somewhat laterally compressed, distally sulcate, proximally tuberculate when dry, glabrous, maroon or dark red or black when mature; pyrenes (25–) 30–60 in 6 or 8 radiating double-files, often of unequal length, subvertically oriented.

Timonius pubistipulus resembles *T. belensis*, but is readily distinguishable by its stipules, which are densely pale-sericeous on the adaxial surface, and its subvertically (rather than obliquely) oriented pyrenes. In *T. trichocladus* and *T. amungwiwanensis,* the pyrenes are also subvertical, but the stipules are adaxially glabrous; in *T. trichocladus* the corollas of pistillate flowers are 4-lobed; in *T. amungwiwanensis* the leaves are pubescent on the abaxial costa. Adaxially pubescent stipules are also found in *T. longitubus* from the Solomon Islands and in *T. oktediensis* from central New Guinea, but in those species the pubescence is much less dense and not consistently present in all collections.

Timonius pubistipulus is reasonable well collected, and most specimens can be assigned with confidence to one of two groups, depending upon presence or absence of branchlet pubescence. The resulting taxa are largely sympatric in New Guinea, but I am hesitant to recognize them as distinct species on the basis of the pubescence character alone and assign them varietal status.

Key to the Varieties of Timonius pubistipulus

1. Branchlets glabrous except above stipule scars. 21a. *T. pubistipulus* var. *pubistipulus*.
1. Branchlets conspicuously and persistently pubescent. 21b. *T. pubistipulus* var. *pubescens*.

21a. Timonius pubistipulus var. pubistipulus.

Bark gray and smooth to brownish and pustular or scaly, the wood reportedly white to reddish straw-colored. Branchlets glabrous. Stipules to 80 mm long, often reddish and conspicuous. Petioles 4–20 mm long; leaf blades (3.5–) 5–12 cm long, (1.5–) 3–5.5 cm wide. Corollas of staminate and pistillate flowers white to pinkish with white lobes. Fruits 6–9 mm long, (7–) 8–10 mm in diameter, dark red or black when mature; pyrenes 25–60. Fig. 6m.

Phenology. Flowering and fruiting throughout the year.

Distribution (Fig. 19). Eastern New Guinea; montane *Araucaria*, *Nothofagus*, and moss-forests or swamps, and frequently at forest edges, in burned-over areas, and other open habitats; (350–) 1500–3400 m.

Vernacular names. *Komai* (Kiliga), *Kormai* (Enga), *Kumai* (Hagen, Kiliga, Togoba), *Kume* (Mendi), *Kumeh* (Onim), *Ome* (Mendi); *Haikapula* (Ialibu); *Abalan* (Enga); *Moghon* (Nerenavip); *Sikil* (Togoba); *Takuneh* (Naho); *Torobinkama* (Sinasina-Nimai); *Tombarumb*, *Tombumpf*, or *Tomburump* (Minj); *Yambasi* (Fiyugi).

ADDITIONAL SPECIMENS EXAMINED. **Papua New Guinea.** [*Northeastern New Guinea.*] WEST SEPIK: S of Nerenavip, Hindenburg Range, *NGF 32144* ♀ (A, BRI, CANB, L, LAE), *NGF 32181* ♂ (A, BRI, CANB, K, L, LAE).—MADANG: south slopes Finisterre Range to S of Mt. Abilala, *Pullen 6132* ♀ (BM, CANB, L, LAE).—MOROBE: Mannasat, Cromwell Mts, *Hoogland 9600* ♀ (A, BM, BRI, CANB, K, L, LAE); Sarawaket, *Clemens & Clemens 5758* ♀ (A); summit of Mt. Shongul, *NGF 37434* ♂ (L, LAE); New Yamap, *NGF 27552* ♀ (A, BRI, CANB, K, L, LAE, SING); Aseki Rd, ca. 57 km from Bulolo, *LAE 75226* ♀ (L, LAE, NO, UPNG); Angabena ridge near Haumuga, Aseki area, *Schodde & Craven 4743* ♀ (A, BRI, CANB, K, L, LAE); Edie Creek rd above Wau, *NGF 13489* ♀ (BRI, CANB, LAE); Bulldog Rd ca. 20 km S of Wau, *Fallen 550* (L, LAE); above Bakaia, ca. 15 mi SW of Garaina, *Hartley 12716* ♀ (A, BRI, CANB, K, L, LAE, RSA).—WESTERN HIGHLANDS: Sirunki, Marapamanda, E of Laigam and N of Yamara, *ANU 762* ♀ (A, CANB, K, L, LAE); N slopes of Sugarloaf complex along Wapu River, *Hoogland & Schodde 7089* ♀ (A, BM, BRI, CANB, K, L, LAE, Z); ca. 3 mi SE of Lake Iviva, *Saunders 1035* (CANB, LAE); ca. 5 mi N of Londoli, Ambum Valley on Wabag–Maramuni Rd, *Saunders 1013* ♂ (CANB, LAE); River Kum, Mt. Kum, *LAE 50320* ♀ (A, CANB, K, L, LAE); Tomba, *NGF 35512* ♂ & ♀ (LAE); top of Mt. Oga, *Saunders 677* (A, BM, L, LAE); slopes of Mt. Kinkain, Kubor Range, *Saunders 741* (CANB); Nebilyer Logging Area, 28 km WNW of Mt. Hagen, *Streimann 8424* ♀ (A); upper Nona River, *Saunders 747* (A, BM, BRI, CANB, L, LAE), *Saunders 748* (CANB), *Saunders 753* ♀ (A, BM, CANB, L, LAE); Minj-Nona Divide, N side of Kubor Range, *Pullen 5167* ♀ (A, BRI, CANB, K, L, LAE); Kubor Range above Kuli, *Robbins 524* ♀ (L, LAE).—EASTERN HIGHLANDS: 2 mi SW of Koge Mission, Sinasina, *Hide 309* ♀ (LAE), *Hide 486* (LAE).—[*Papua.*] SOUTHERN HIGHLANDS: Mimimbipl, 20 mi from Mendi, Tambul Rd, *LAE 55321* ♀ (A, BISH, K, L, LAE); western slope of Mt. Giluwe above Klareg, *Schodde 2015* ♀ (A, BRI, CANB, K, L, LAE); SW slope of Mt. Giluwe, *Schodde 1673* ♂ (A, CANB, L, LAE); base of Mt. Giluwe, *LAE 74437* ♀ (BRI, CANB, L, LAE); slopes of Mt. Giluwe, *NGF 32502* ♀ (A, BRI, CANB, E, K, L, LAE, SING); Omin, *ANU 20326* (LAE), *ANU 20353* (LAE); Loma Sawmill, 6 km SE of Ialibu, *Vinas 60* ♀ (A).—CENTRAL: E of Mt. Tufa, *Brass 4077* ♀ (BRI); Mt. Tufa, *Brass 5046* ♀ (A, BO, BRI, K, L); rd from Woitape to Kosipi, Iowasi Swamp, *NGF 20308* ♀ (A, BO, BRI, CANB, LAE, SING); E slope Lake Myola no. 2, *NGF 34629* ♀ (A, BISH, E, K, L, LAE, M, QRS); Koitaki, *Carr 12522* ♀ (A, BM, CANB, K, L, SING); Boridi, *Carr 13095* ♀ (A, CANB, L, SING), *Carr 13343* ♀ (A, BM, CANB, K, L, SING), *Carr 14607* ♂ (A, BM, CANB, K, L); Lala Valley, *Carr 15856* (BM, L, SING); The Gap, *Carr 15042* ♂ (A, BM, L, SING); above The Gap, *Carr 13819* ♀ (A, BM, CANB, K, L); Sogeri region, *Forbes 770* ♀ (A, BM, E, L).—NORTHERN: Kokoda, *Carr 16453* ♀ (B, CANB, L), *Carr 16478* ♂ (CANB, G, K, L); E side of Lake Myola no. 1, *LAE 61972* ♂ (A, BISH, BO, CANB, K, L, LAE, M, MO, QRS, SING).—MILNE BAY: junction Ugat and Mayu rivers near Mayu 1, *NGF 28897* ♀ (A, BISH, CANB, E, K, L, LAE, SING); Mayu 2, Mt. Suckling, *LAE 54117* ♀ (A, CANB, K, L, LAE); Goropu Mts, Pumpunipon to S ridge of Goe Dendeniwa, *Veldkamp & Stevens 5694* ♂ (BISH, L, LAE); track, S spur to Pumpunipon, Mt. Suckling complex, *LAE 54178* ♀ (A, CANB, L, LAE); N slopes of Mt. Dayman, Maneau Range, *Brass 22304* ♀ (A, CANB, L, LAE), *Brass 22702* ♀ (A, CANB, K, L, LAE), *Brass 22874* ♀ (A, CANB, L, LAE); Mt. Dayman, *Paijmans 1090* ♀ (CANB); Mt. Wadimana, NE outlier of Mt. Simpson, *Pullen 7842* ♂ (A, L, LAE); summit area, Mt. Wadimana, *Schodde 5472* ♂ (CANB, L, LAE); Aparamu Ridge, E of Mt. Simpson Range, *Schodde 5519* (CANB).

The bark varies from gray and smooth to brownish and pustular or scaly; the wood is reported as white to reddish straw-colored. Stipules, leaves, and fruits are often somewhat larger than in var. *pubescens*, but the branchlet pubescence is the only reliable character separating the taxa.

21b. Timonius pubistipulus var. **pubescens** S. Darwin, J. Arnold Arb. 64: 614. 1983.—TYPE: PAPUA NEW GUINEA. Morobe: Mt. Kaindi near Edie Creek, heavy liana in a *Nothofagus* forest, 2400 m, 9 May 1963, *Hartley 11840* ♀ (holotype: A!; isotypes: C! CANB! G! L! LAE!).

Bark pale gray to dark brown and smooth, the wood reportedly white to pale brown. Branchlets usually conspicuously and persistently pubescent with stiff, spreading hairs. Stipules to 50 (rarely to 60) mm long. Petioles (2–) 5–10 mm long; leaf blades 2.5–8 (–12) cm long, 1–3.5 (–5) cm wide. Corollas of staminate and pistillate flowers yellowish to brownish white or rarely dull pinkish. Fruits 5–8 mm long, 6–8 mm in diameter, maroon or black when mature; pyrenes 40–60. Fig. 8h, i.

Phenology. Flowering and fruiting throughout the year.

Distribution (Fig. 19). Snow Mountains eastward to the Owen Stanley Range of southeastern New Guinea; montane forests, usually in swamps, myrtaceous scrub, edges of trails, ridge-tops, old landslides, and other open or successional habitats; 1470–2600 m.

Vernacular Name. *Komai* (Enga).

ADDITIONAL SPECIMENS EXAMINED. **Indonesia.** IRIAN JAYA [*Western New Guinea*]. Snow Mountains: Bijenkorfbivak, Helling Hellwigberg, *Pulle 773* (A, L). **Papua New Guinea.** [*Northeastern New Guinea.*] WEST SEPIK: Folongonom, second bush camp below Tamanagabip on track to Busilim, *LAE 59483* ♀ (A, BISH, E, K, L, LAE, M).—MOROBE: NE slope of Mt. Rawlison, *Hoogland & Craven 9341* ♀ (A, BM, BRI, CANB, K, L, LAE); Ogeramnang, *Clemens & Clemens 4649* ♀ (A, B); Aseki-Spreader Divide, *LAE 54772* ♀ (A, BISH, CANB, K, L, LAE); E slope Spreader Divide, ca. 7 mi NW of Aseki, *Schodde & Craven 4904* ♀ (A, BRI, CANB, K, L, LAE); near Wengomanga, via Oiwa, *Craven & Schodde 1291* ♀ (CANB, LAE); along water race of N. G. Mine, *NGF 21063* ♀ (A, BISH, BRI, CANB, E, K, L, LAE, M, SING); Edie Creek, *NGF 19948* (CANB, L, LAE), *NGF 19959* ♀ (A, BISH, BO, BRI, CANB, E, L, LAE, SING), *Hartley 11725* ♂ (A, BRI, CANB, K, L, LAE, RSA); Mt. Kaindi, SE slope, *Allison NG-195* ♀ (LAE); track descending SE from S summit of Mt. Kaindi, *UPNG 3114* (UPNG); ca. 14 km WSW of Wau, Mt. Kaindi, *Allison NG-171* ♀ (LAE); Mt. Kaindi summit, *NGF 13908* ♀ (A, BO, BRI, CANB, K, L, LAE), *LAE 77669* ♂ (LAE, NO), *LAE 77670* ♀ (A, LAE, NO); Kaindi, *Brass 29559* ♀ (A, BO, CANB, K, L, LAE), *Brass 29561* ♂ (A, CANB, K, L, LAE); ca. 15 km SSW of Wau, Bulldog Rd, *WEI 1374* ♀ (LAE); Bulldog Rd, 12 km SSE of Wau, *Pratt NG-105* ♀ (LAE); Bulldog Rd ca. 13 km SSE of Wau, *Allison NG-161* ♀ (LAE); Lake Trist, *NGF 29135* ♀ (A, BRI, CANB, K, L, LAE, SING); 5 km E of Lake Trist, *Paijmans 1568* ♀ (CANB, LAE).—WESTERN HIGHLANDS: top of ridge 3 mi S of Kompiam, *ANU 2805* ♀ (A, CANB, K, L, LAE).—EASTERN HIGHLANDS: confluent Warapuri and Kori rivers, Wahgi–Jimmy Divide, N of Nondugl, *NGF 18173* ♀ (BRI, CANB, L, LAE); summit Mt. Elandora, *Jermy 5025* ♀ (LAE).—[*Papua.*] SOUTHERN HIGHLANDS: Munie Logging Area, 14 km NW of Ialibu, *Streimann 8531* ♀ (A); Kengaput, 6 km SSE of Mendi, *Vinas 95* ♀ (A); Onim, *Rau 160* ♀ (A, K, L, LAE).—WESTERN: Mt. Bosavi, N side, *Jacobs 8859* ♀ (BISH, L, LAE).

The bark is pale gray to dark brown and smooth; the wood is reported as white to pale brown. On the basis of available collections, it appears that var. *pubescens* extends farther west but not so far east as var. *pubistipulus*, but the varieties are sympatric through most of their range and occupy similar habitats.

22. Timonius scabriflorus (Valeton) Merrill & Perry, J. Arnold Arb. 26: 238. 1945. *Timonius avenis* var. *scabriflorus* Valeton, Bot. Jahrb. Syst. 61: 37 ('scabriflora'). 1927.—TYPE: [PAPUA NEW GUINEA.] Northeastern New Guinea. Madang: ca. 20 km SW of Bulu ("Kani-Gebirge"), in forest at ca. 1000 m, 24 Mar 1908, *Schlechter 17504* ♀ (lectotype, here designated: L!; isolectotypes: A! BM! BRI! C! G! S!)

Epiphytic shrubs or climbers or trees to 17 m tall and 15 cm dbh. Bark gray, sometimes mottled white, smooth, the wood white to yellowish, soft or hard. Branchlets 2–3 mm wide toward apex, brown to gray or black, densely and finely puberulent with spreading hairs. Stipules to 40 mm long and 15 mm wide, ovate, acuminate, densely and finely whitish canescent abaxially but with longer, spreading, brownish hairs toward costa and base, glabrous adaxially except for colleters toward base. Petioles 2–10 (–15) mm long, 2–3 mm wide at middle, semi- to subterete, puberulent; leaf blades 3–11 (–14) cm long, 1–4.5 (–6) cm wide, elliptic to narrowly ovate to obovate, acuminate at apex, acute to attenuate to abruptly rounded at base, coriaceous, glabrous above except often ciliate toward apex or base, pubescent with scattered, spreading to subappressed hairs over lamina or more densely pubescent on costa beneath, the costa plane to canaliculate above, elevated beneath, the secondary nerves hardly distinguishable, spreading, the veinlets prominulous above, paxillate; domatia pitlike, distributed along costa or over lamina beneath, pubescent, often absent. Staminate inflorescences 3- to 5- (7-) flowered, at anthesis to 1.8 cm long and 0.5 cm wide; peduncles 3–8 mm long, 1–2 mm wide, puberulent; bracts absent. Staminate flowers: calyx limb cupular, truncate to undulate, glabrous, the tube 0.5–1 mm long from base; corolla hypocrateriform to narrowly infundibular, white to greenish, the tube 8–10 mm long, ca. 1.5 mm wide, densely spreading-puberulent outside, glabrous within, the lobes 4, to 3 mm long, 1.5 mm wide, acute, the corolla limb 6–8 mm in diameter; stamens 4, the anthers 5–6 mm long, ca. 0.5 mm wide, linear-oblong, glabrous; pistillodium ca. 3 mm long, filiform, glabrous. Pistillate inflorescences 1-flowered; peduncles 2–7 mm long, 2–3 mm wide, puberulent; bracts absent. Pistillate flowers: calyx limb cupular, truncate to undulate, glabrous, the tube ca. 0.5 mm long; corolla narrowly infundibular, white to greenish, the tube (late bud) 5 mm long, 2 mm wide, densely spreading-puberulent outside, glabrous within, expanded corolla limb not seen; staminodia not seen; hypanthium 2.5–3 mm long, ca. 3 mm in diameter, depressed-cylindric to subglobose, glabrous or rarely scattered-puberulent, the disc glabrous, the style not seen. Fruits 5–6 mm long, 7–9 mm wide, subglobose, conspicuously tuberculate-sulcate when dry, glabrous, color of mature fruits unknown; pyrenes 60–100 in 8 radiating double-files, subequal, obliquely oriented.

Phenology. Collected in flower and fruit from March to November, but likely flowering and fruiting throughout the year.

Distribution (Fig. 19). Northeastern New Guinea; in primary or secondary forests and often associated with *Castanopsis*; 1000–2000 m.

Vernacular Name. *Hondai* (Habi'im).

ADDITIONAL SPECIMENS EXAMINED. **Papua New Guinea.** [*Northeastern New Guinea.*] MOROBE: Yunzaing, *Clemens & Clemens 3568bis* ♀ (A, Z), *Clemens & Clemens 3586a* ♂ (A, BR, BRI, Z), *Clemens & Clemens 4143* ♀ (BR, G, Z); ridge E of Poverty Creek, 8 km NE of Wau, *Pratt NG-3* ♀ (LAE).—EASTERN HIGHLANDS: Arau, *Brass 32230* ♂ (A, CANB, L, LAE); Kainantu-Okapa rd, 10 mi W of Kainantu, *Hartley 12430* ♀ (A, BRI, CANB, L, LAE, RSA); Habi'ina, near To'okena, Obura, slopes of Mt. Piora, *Coode 3885* ♀ (LAE); 10 mi SE of Obura, *Hays 269* ♀ (LAE), *Hays 363* ♀ (LAE).

Timonius scabriflorus is similar to *T. belensis*, but has young parts and inflorescences more densely spreading-pubescent, as are the lower surfaces of leaf blades. *Schlechter 17504* was the only collection cited by Valeton, the holotype presumably deposited at B and no longer extant; among the duplicates examined, the one located at Leiden is most complete and is here designated the lectotype.

23. Timonius secundiflorus S. Darwin, sp. nov.—TYPE: PAPUA NEW GUINEA. West Sepik: Torricelli Mts, ca. 4 mi N of Wigote, ca. 1520 m, 17 Sep 1961,

Darbyshire 455 ♀ (holotype: A!; isotypes: B! BISH! BRI! CANB! G! K! L! LAE!).

Timonius avenis f. *triflora* Valeton, Bot. Jahrb. Syst. 61: 38. 1927.—TYPE: [PAPUA NEW GUINEA.] Northeastern New Guinea. Aprilfluss, Lager 18, 165 m, Nov 1912, *Ledermann 9853* ♀ (lectotype, here designated: L!).

Timonius ex affinitate T. avenis Valeton sed ab fructibus subglobosis (non tuberculatis) saepe glaucis pyrenis oblique dispositis differt.

Epiphytic shrubs to seemingly scandent trees to 17 m tall and 30 cm dbh. Bark smooth, reddish brown to gray, the wood straw-colored. Branchlets 1–2 mm wide toward apex, pale, minutely canescent to subglabrous. Stipules to 35 mm long and 8 mm wide, lanceolate, acuminate, glabrous abaxially, glabrous adaxially except for colleters toward base. Petioles to 10 mm long, less than 1 mm wide at middle, semi- to subterete, essentially glabrous; leaf blades 2–7 cm long, 0.5–3 cm wide, narrowly elliptic, sharply long-acuminate at apex, attenuate at base, coriaceous, glabrous above, glabrous or sometimes subciliate at margin toward base beneath, the costa plane to canaliculate above, elevated beneath, the secondary nerves indistinct, the veinlets prominulous above, paxillate; domatia pocketlike on lamina near costa, and sometimes also pitlike and scattered over lamina beneath, subglabrous. Staminate inflorescences (3-) 7- to 11-flowered, at anthesis to 2.5 cm long, ca. 2 cm wide; peduncles 5–10 mm long, less than 1 mm wide, glabrous, often glaucous; bracts absent. Staminate flowers: calyx limb cupular-cylindric, truncate to shallowly 4-dentate, glabrous, usually glaucous, the tube 1.5–2.5 mm long from base; corolla narrowly infundibular, greenish outside, white within, the tube to 8 mm long, 0.5–1 mm wide, glabrous outside, glabrous within, the lobes 4, ca. 2 mm long, 1 mm wide, lanceolate, acuminate, papillose, the corolla limb to 5 mm in diameter; stamens 4, the filaments essentially absent, the anthers 2.5–3 mm long, ca. 0.5 mm wide, oblong, glabrous; pistillodium not seen. Pistillate inflorescences 1-flowered; peduncles 2–10 (in fruit to 15) mm long, ca. 1 mm wide, glabrous; bracts absent. Pistillate flowers: calyx limb cupular-cylindric, truncate, glabrous, the tube 1–1.5 mm long; corolla narrowly infundibular, greenish to white, caducous, the tube 5–6 mm long, ca. 1.5 mm wide, glabrous outside, glabrous within, the lobes 4–8, 2–2.5 mm long, ca. 1 mm wide, ovate, acute, the corolla limb ca. 7 mm in diameter; staminodia 4–8, ca. 1.5 mm long, 0.5 mm wide, sagittate, glabrous; hypanthium to 3 mm long, to 4 mm in diameter, subglobose, glabrous, usually glaucous, the disc glabrous, the style ca. 6 mm long, ca. 0.5 mm wide near middle, 6- to 8-armed, glabrous. Fruits 6–8 mm long, 5–10 mm in diameter, globose to depressed-globose, not at all costate or tuberculate when dry, glabrous, often glaucous, color of mature fruits unknown; pyrenes 35–80 in 8 radiating double-files, subequal, obliquely oriented. Figs. 4c, 6k, l, 21.

Phenology. Collected in flower from March to October, in fruit more or less throughout the year

Distribution (Fig. 19). North-central and northeastern New Guinea; montane forest, ridge forest, and mossy forest; 400–2500 m.

Vernacular Name. *Favell* (Wapi).

ADDITIONAL SPECIMENS EXAMINED. **Indonesia.** IRIAN JAYA [*Western New Guinea*]. Djajapura: 4 km SW of Bernhard Camp, Idenburg [Taritatu] River, *Brass 13310* ♀ (A, BM, BO, BRI, L, LAE), *Brass*

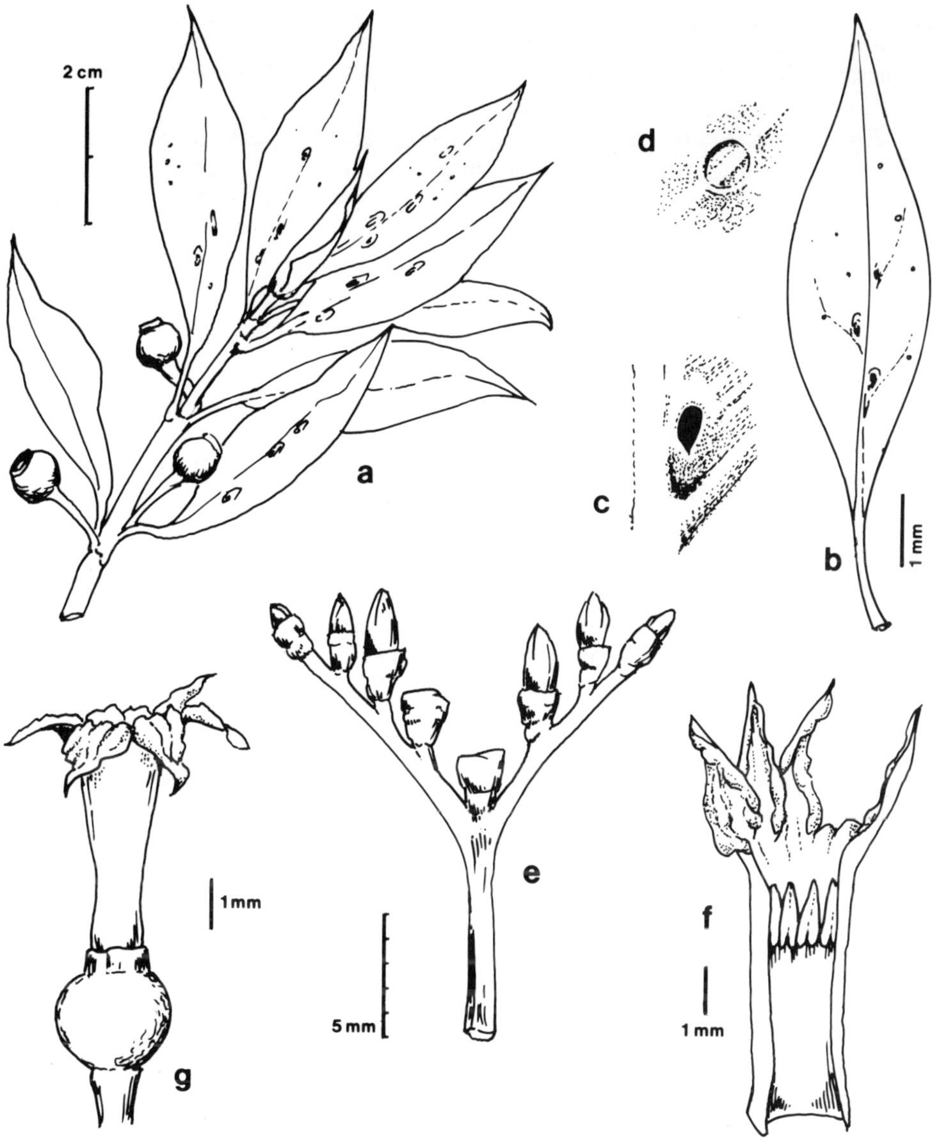

FIG. 21. *Timonius secundiflorus*. a. Branchlet with fruits. b. Abaxial view of leaf. c. Same, showing domatium in axil of secondary nerve arising from basal region of abaxial costa (magnified). d. Same, showing domatium associated with tertiary nerve from distal part of abaxial lamina (magnified). e. Staminate inflorescence. f. Corolla of pistillate flower, opened to show staminodia. g. Pistillate flower. (Based on: a–d, f, *Darbyshire 455*; e, *Hoogland & Craven 10950*.)

13628 ♀ (A, BM, BO, BRI, L, LAE); 6 km SW of Bernhard Camp, Idenburg [Taritatu] River, *Brass 12913* ♀ (A, BM, BO, BRI, L, LAE); Hollandia, Cycloop Mts, Ifar–Ormoe trail, *van Royen & Sleumer 5900* ♀ (A, BRI, G, K, L, LAE); Cycloop Mts, path to Mt. Rara, *van Royen & Sleumer 5939* ♂ (A, L, LAE). **Papua New Guinea.** [*Northeastern New Guinea.*] West Sepik: Prospect Creek near Frieda River, *NGF 42562* ♀ (A, BISH, BRI, CANB, K, L, LAE, SING).—East Sepik: Etappenberg, *Ledermann 8899* (SING); Lordberg, *Ledermann 9904* (L), *Ledermann 9945* (L), *Ledermann 10292* (L); Mt. Hunstein, *Ledermann 8426* ♀ (SING), *Ledermann 8479* ♂ (L, SING); summit of Sumset (Mt.

Hunstein), *Hoogland & Craven 10986* ♀ (A, BRI, CANB, K, L, LAE); E ridge of Sumset (Mt. Hunstein), *Hoogland & Craven 10950* ♂ (A, BRI, CANB, G, K, L, LAE).—EASTERN HIGHLANDS: Marafunga, ca. 20 mi NW of Goroka, *Hartley 13222* ♀ (A, BRI, C, CANB, G, K, L, LAE, RSA).

Timonius secundiflorus is easily recognized by the usually glaucous inflorescences, often secund staminate flowers, and subglobose fruits with mesocarp that usually does not shrink against the pyrenes when dried. It resembles *T. avenis*, but the pyrenes are obliquely rather than vertically arranged and the leaf domatia are much more prominent. In publishing *Timonius avenis* f. *triflora*, Valeton cited five collections from Northeastern New Guinea, but only *Ledermann 9853* appears to be extant and is here designated the lectotype.

24. Timonius singularis (F. v. Mueller) L. S. Smith, Proc. Roy. Soc. Queensland 68: 50. 1957. *Abbottia singularis* F. v. Mueller, Fragm. 9: 181. 1875.—TYPE: AUSTRALIA. Queensland: mountains around Rockingham Bay, *Dallachy s.n.* (holotype: MEL! ♂ & ♀; isotypes?: BO! M!).

Lithophilic shrubs, epiphytes, or trees to 25 m tall and 80 cm dbh. Bark brownish, the underbark brownish or speckled pink and brown; nature of wood unknown. Branchlets 2–3 mm wide toward apex, pale, at length becoming 4-angled or terete, subglabrous. Stipules to 60 mm long and 8 mm wide, long-lanceolate to linear, acuminate, glabrous abaxially, glabrous adaxially except for colleters near base. Petioles 10–30 mm long, 1–2 mm wide at middle, semiterete to canaliculate, glabrous; leaf blades 4–14 cm long, 2–6 cm wide, elliptic to somewhat oblanceolate, sharply acuminate at apex, attenuate at base, thick-chartaceous to coriaceous, glabrous above, glabrous beneath, the costa prominulous to canaliculate above, elevated beneath, the secondary nerves hardly distinguishable, the veinlets prominulous above, submerged beneath, paxillate; domatia absent or occasionally open and pocketlike on lamina, or distributed along costa and cryptlike with slitlike openings, glabrous to sparsely pubescent. Staminate inflorescences 3- to 7-flowered, at anthesis 1.5–2 cm long, 0.5–1.5 cm wide; peduncles 5–6 mm long, 1–2 mm wide, compressed, glabrous; bracts absent. Staminate flowers: calyx limb cupular, truncate to undulate, glabrous, the tube 2–3 mm long from base; corolla narrowly infundibular, white to greenish, the tube 8–12 mm long, ca. 1 mm wide, glabrous outside, glabrous within, the lobes 4, 3–5 mm long, 1–1.5 mm wide, lanceolate, acute, the corolla limb to 10 mm in diameter; stamens 4, the anthers ca. 3.5 mm long, linear-sagittate, glabrous; pistillodium not seen. Pistillate inflorescences 1-flowered; peduncles ca. 5 mm long (to 10 mm long in fruit), 1–2 mm wide, compressed, glabrous; bracts absent. Pistillate flowers: calyx limb cupular, truncate, glabrous, the tube ca. 1 mm long; corolla infundibular, white to greenish, the tube 4–5 mm long, 2–3 mm wide, glabrous outside, glabrous within, the lobes 8–11, to 5 mm long, 1 mm wide, lanceolate to subulate, acute, glabrous, the corolla limb to 12 mm in diameter; staminodia 8–11, 2–2.5 mm long, ca. 0.5 mm wide, oblong-sagittate, glabrous; hypanthium 2–3 mm long, ca. 4 mm in diameter, depressed-globose, glabrous, the style 6–8 mm long, 8- to 11-armed, glabrous. Fruits 5–6 mm long, 8–10 mm in diameter, depressed- and somewhat compressed-globose, proximally tuberculate, glabrous, black when mature; pyrenes 35–90 in 8–12 radiating double-files, subequal, obliquely oriented. Fig. 6h, i.

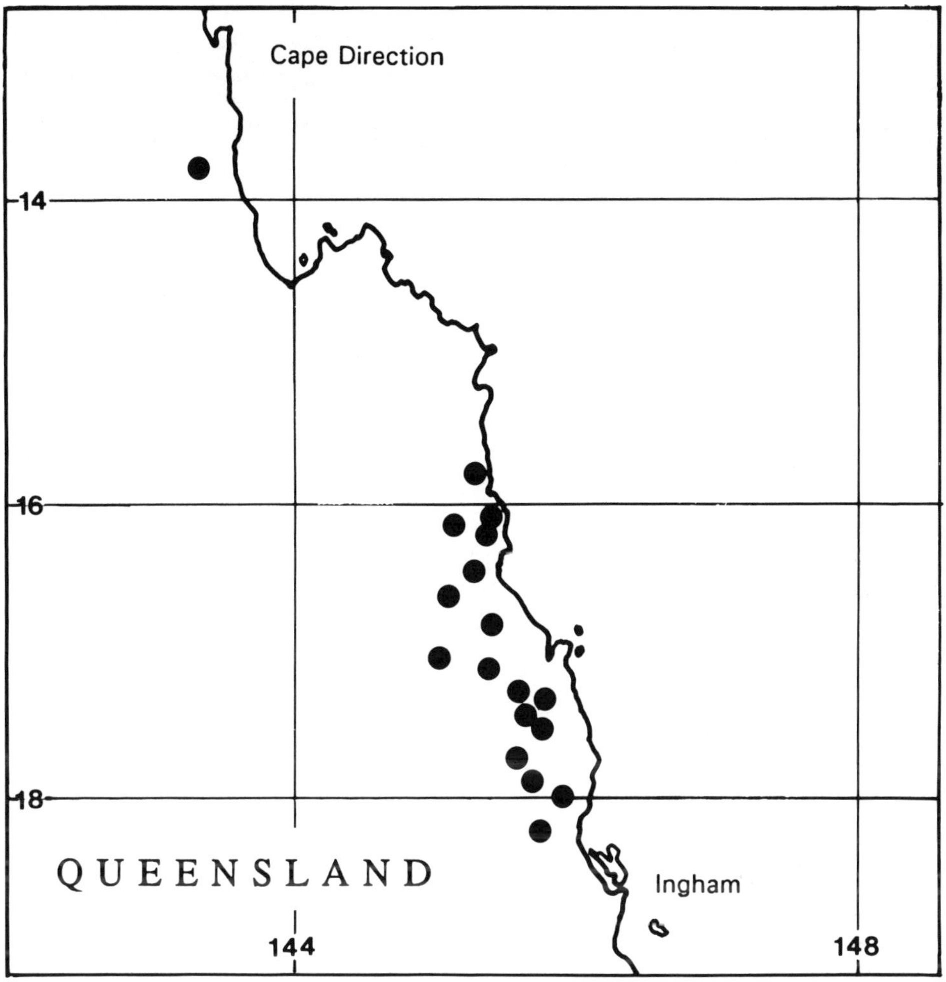

FIG. 22. Distribution of *Timonius singularis*.

Phenology. Flowering and fruiting throughout the year.

Distribution (Fig. 22). Australia, eastern Queensland (Cape York Peninsula) from McIlwraith Range southward to Rockingham Bay; on granitic outcrops and ridgetops, or in rain forests, mesophyll vine-forests, and microphyll mossy thickets; 400–1300 m.

Vernacular names. *Mountain-mangrove, False fig.*

ADDITIONAL SPECIMENS EXAMINED. **Australia.** QUEENSLAND: McIlwraith Range, *Hyland 7621* ♂ (BRI, CANB, L, QRS); Mt. Finnegan, *Brass 20338* ♀ (A, BRI, CANB, L, SING); summit of Mt. Finnegan, *Webb & Tracey 10845* ♀ (BRI); Mt. Hemmant, N of Noah Creek, Cape Tribulation area, *Webb & Tracey 11711* ♀ (BRI, CANB); Mt. Hemmant, between Daintree River & Cape Tribulation, *Tracey 14071* ♀ (BRI, QRS); Hilda Creek, *Irvine 2284* ♀ (QRS); Thornton Peak, *L. S. Smith 208* ♂ (BRI), *Brass 2295* ♂ (A, BISH, BRI); Windsor Tableland State Forest Reserve 144, *Crome 884* ♀ (NO, QRS); summit of Mt. Spurgeon, *Stocker 775* ♀ (BRI, L, QRS); Rock Site, N of junction Platypus Creek & Mossman River, *Webb & Tracey 11335* (BRI); Platypus Creek at head of Mossman River, *Tracey*

14895 (QRS); State Forest Reserve 143, South Mary logging area, *Hyland 11440* ♀ (NO, QRS); State Forest Reserve 143, North Mary logging area, *Hyland 7020* (BRI, QRS), *B. Gray s.n.* ♂ (QRS); State Forest Reserve 143, Riflemead, Carbine logging area, *Hyland 13596* ♀ (QRS); Mt. Lewis Road, 28 km from Mt. Molloy–Mossman rd, *Jessup & Clarkson 290* ♀ (BRI, CANB); Mt. Edith, 6.4 km N of Danbulla, *L. S. Smith 3755* ♀ (A, BRI, L, RSA); State Forest Reserve 185, Breach logging area, *Hyland 5235* ♀ (QRS); Gadgarra Reserve, Peeramon, Atherton Tableland, *Kajewski 1191* ♀ (A, BRI); State Forest Reserve 185, Outlook Rock, *Irvine 819* ♂ & ♀ (BRI, L, QRS); Boonjee, near Malanda, *S. T. Blake 15220* ♀ (QRS); State Forest Reserve 755, ca. 17 km ESE of Malanda, *Moriarty 2028* ♀ (BRI, CANB, L, QRS); State Forest Reserve 755, North Johnstone logging area, *Dockrill 1434* ♀ (QRS); Koolmoon Creek, 17.6 km SSE of Ravenshoe, *L. S. Smith 4705* ♀ (BRI, L); State Forest Reserve 756, Maple logging area, between Tully River & South Johnstone River, *Dockrill 33* ♂ (A, BISH, BRI, CANB, L, QRS); State Forest Reserve 756, Karang logging area, 48 mi S of Cairns, *Hyland 3583* ♂ (A, BRI, L); State Forest Reserve 756, *Risley 96* ♀ (CANB, L, QRS).

Timonius singularis is the only species of subg. *Abbottia* present in Australia. It is confined to the coastal ranges of tropical Queensland, where it is often an epiphyte on species of *Ficus*, *Eugenia*, or *Endiandra*, or terrestrial. *Timonius singularis* resembles *T. affinis*, *T. belensis*, and such Papuan species as *T. meridionalis*, and is distinguishable in having relatively short peduncles, pistillate corollas, and hypanthia. Leaf domatia are often present, but were not found in specimens collected north of Port Douglas. Among the possible isotypes cited above, the specimen at Bogor includes both staminate and pistillate material and bears a Melbourne Museum label but not the name of the collector; the Dallachy specimen at Munich, also with a Melbourne label, consists of just a few pistillate fragments.

25. Timonius subavenis (Valeton) S. Darwin, J. Arnold Arb. 64: 616. 1983. *Timonius scaber* var. *subavenis* Valeton, Bot. Jahrb. Syst. 61: 51. 1927.—Type: [Papua New Guinea.] Northeastern New Guinea. "Sepik-Gebiet" [vicinity of Malu and April River, 3–23 Sep 1912], *Ledermann 8793* ♂ (lectotype, designated by Darwin, 1983: SING!).

Epiphytic shrubs or, at maturity, small trees; nature of bark and wood unknown. Branchlets 2–3 mm wide toward apex, brown to dark gray-brown, shaggy-pubescent with brownish hairs. Stipules to 45 mm long and 10 mm wide, lanceolate, acuminate, shaggy-hirsute toward center and base abaxially, glabrous adaxially except for colleters near base. Petioles 5–15 mm long, 1–2 mm wide at middle, semi- to subterete, hirsute; leaf blades 4–14 cm long, 1.5–5 cm wide, long-elliptic to oblanceolate, long-acuminate to sharply acute at apex, attenuate at base, thin-coriaceous to chartaceous, glabrous to subciliate toward apex or base to scattered-puberulent over lamina above, scattered appressed- to spreading-pubescent with brownish hairs over lamina and more densely pubescent on costa beneath, the costa plane to canaliculate above, elevated beneath, the secondary nerves obscure, spreading, the veinlets discernible above, paxillate; domatia absent. Staminate inflorescences 3- to 11-flowered, at anthesis 1.5–3 cm long, 1.5–2.5 cm wide; peduncles 2–8 mm long, ca. 1.5 mm wide, glabrous to densely tomentose; calyx tube cupular, truncate, distally glabrous, pubescent toward base, the tube 0.5 (–1) mm long from base; corolla narrowly infundibular, white to reddish, the tube (late bud) ca. 6 mm long, densely hirsute outside, the lobes 4; stamens 4; anthers and pistillodium not seen. Pistillate inflorescences 1-flowered; peduncles 1–5 mm long, ca. 2 mm wide, glabrous or with scattered hairs; calyx tube cupular, truncate, glabrous, the tube ca. 0.5 mm long; corolla narrowly infundibular, white to reddish, the tube 6–7 mm long, 1–1.5 mm wide, hirsute outside, the lobes 6, to 2 mm long, ca. 1 mm wide, acute, papillose, the corolla limb 5–6 mm in diameter; hypanthium 3–3.5 mm long, 3–3.5 mm in diameter, subglobose, pubescent with long, scattered hairs to subglabrous, the disc glabrous. Fruits 4–5 mm long, 3.5–6 mm in diameter, subglobose to somewhat laterally compressed, 6- to 8-ribbed, glabrous, color of mature fruits unknown; pyrenes 40–50 in 6–8 radiating double-files, subequal, vertical. Figs. 3l, 20d.

Phenology. Flowering and fruiting throughout the year.

Distribution (Fig. 23). North-central New Guinea; variously reported as an epiphyte, liana, shrub, or small tree in primary forests and swamp forests; 50–1750 m.

ADDITIONAL SPECIMENS EXAMINED. **Indonesia.** IRIAN JAYA [*Western New Guinea*]. Snow Mountains: Bijenkorfbivak, Zuidhelling Hellwigberg, *Pulle 650* ♂ (A, BO, K, L, SING); Pesechem-vallei, *Pulle 1155* ♂ & ♀ (BO, L). Djajapura: Rouffaer River, *Docters van Leeuwen 10141* ♀ (A, BO, K, L), *Docters van Leeuwen 10172* ♂ (BO, K, L); Boden River, 60 km SE of Sarmi, *BW 8101* ♀ (BR, CANB, L, LAE); Bernhard bivouac, *Meyer Drees 499* ♂ (A, BO, K, L, SING), *Meyer Drees 607* ♂ (BO, L). **Papua New Guinea.** [*Northeastern New Guinea.*] WEST SEPIK: Frieda River, airstrip area, *LAE 73981* ♀ (A, E, K, L, LAE, M).

Timonius subavenis is similar to *T. trichanthus*, but occupies a more westerly range. In *T. trichanthus* the basal portion of the calyx (hypanthium) is glabrous or nearly so (versus pubescent in *T. subavenis*), whereas the calyx tube is pubescent and 1 mm or more long (versus glabrous and ca. 0.5 mm long in *T. subavenis*).

26. Timonius trichanthus Merrill & Perry, J. Arnold Arb. 26: 238, fig. 3C. 1945.—TYPE: [PAPUA NEW GUINEA.] Northeastern New Guinea. Morobe: Matap, 1520–1830 m, 6 Feb 1940, *Clemens & Clemens 11158B* ♀ (holotype: A!; isotype: E!).

Slender trees to 12 m tall and 20 cm dbh or shrubs (sometimes epiphytic) to 8 m tall. Bark gray to brown, lenticellate, the underbark straw- or salmon-colored, the wood white to cream-colored. Branchlets 3–4 mm wide toward apex, dark brown to gray, brown-canescent. Stipules to 50 mm long and 20 mm wide, lanceolate, acuminate, densely shaggy-tomentulose with spreading, reddish to stramineous hairs abaxially, glabrous adaxially except strigose among colleters toward base. Petioles 6–15 mm long, 2–3 mm wide at middle, canaliculate to subterete, puberulent; leaf blades 6–14 cm long, 2.5–6 cm wide, broadly elliptic to oblanceolate, acute to cuspidate to acuminate at apex, acute to obtuse to abruptly rounded at base, coriaceous, puberulent on costa and frequently ciliate toward apex and base, or scattered-puberulent over lamina above, scattered appressed-puberulent to tomentulose with stramineous hairs over lamina, or glabrous except on costa beneath, the costa plane to canaliculate above, elevated beneath, the secondary nerves indistinct, the veinlets submerged in lamina, paxillate; domatia appearing as tufts of hairs along costa beneath. Staminate inflorescences 3- to 9-flowered, at anthesis to 3 cm long and 2 cm wide; peduncles to 10 mm long; bracts absent. Staminate flowers: calyx limb cupular, truncate, densely to sparsely appressed-strigillose, the tube 2–3 mm long from base; corolla narrowly infundibular, greenish

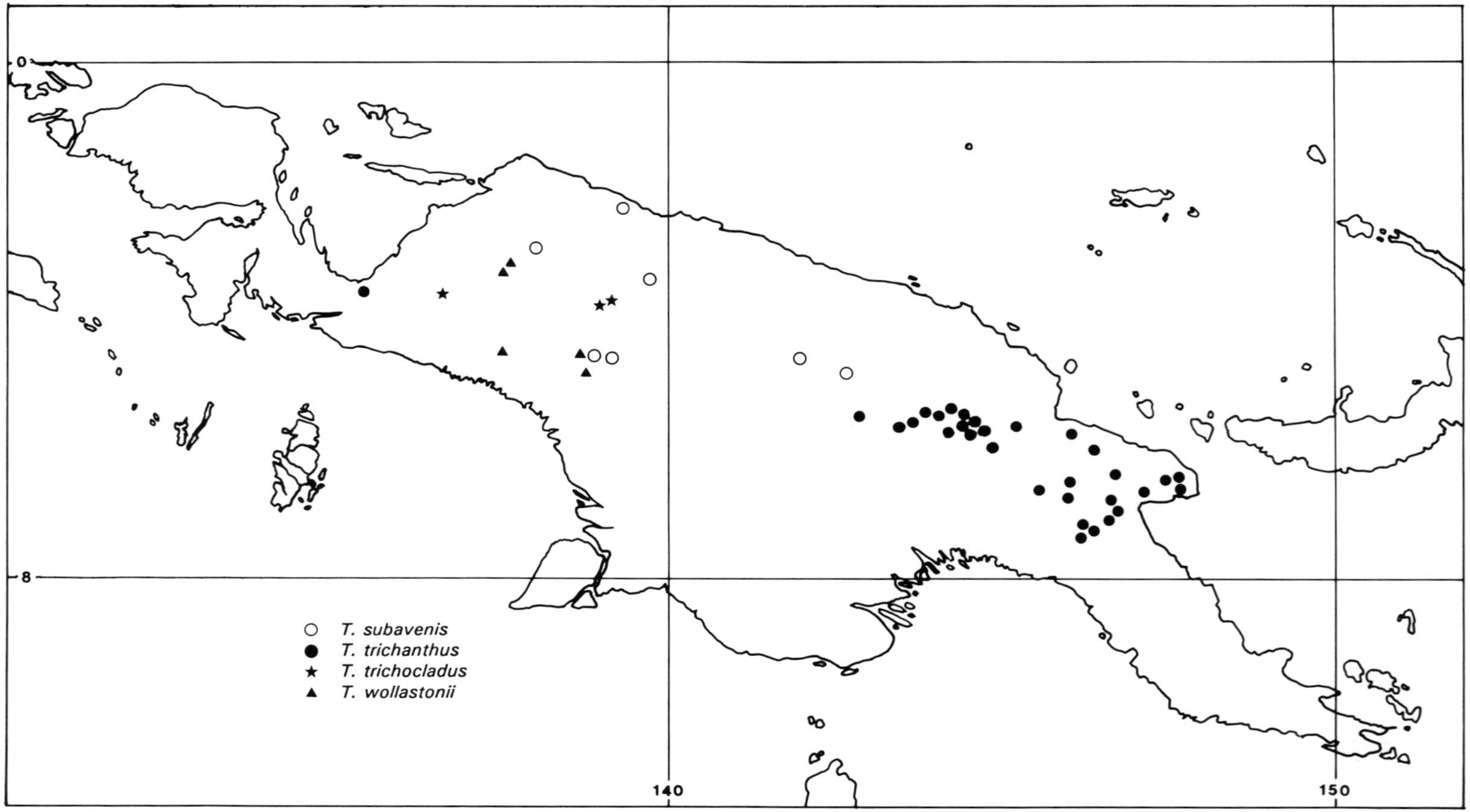

FIG. 23. Distribution of *Timonius subavenis*, *T. trichanthus*, *T. trichocladus*, and *T. wollastonii*.

to white, the tube to 10 mm long, 1–2 mm wide, densely appressed-strigillose outside, glabrous within, the lobes 4 (–6), 1.5–3 mm long, acute, papillose, the corolla limb to 6 mm in diameter; stamens 4 (–6), the filaments 1–1.5 mm long, the anthers to 3 mm long, 0.3 mm wide, oblong, glabrous; pistillodium to 2.5 mm long, filiform. Pistillate inflorescences 1-flowered; peduncles (2–) 10–20 mm long, ca. 1 mm wide, expanding distally, spreading-pubescent; bracts usually absent or to 2 mm long, deltate. Pistillate flowers: calyx limb cupular, truncate, pubescent with long appressed hairs, the tube 1–1.5 mm long; corolla narrowly infundibular, greenish to white, the tube to 7 mm long, to 2.5 mm wide, densely strigillose outside, glabrous within, the lobes 6 (–8), 2–2.5 mm long, narrowly ovate, acute, papillose, the corolla limb 5–8 mm in diameter; staminodia 6, 2–3 mm long, sagittate; hypanthium 2.5–3 mm long, 2.5–4 mm in diameter, subcylindric, glabrous or with a few scattered hairs, the disc subconical to turbinate, glabrous, the style to 9 mm long, to 1 mm wide near middle, 6- (8-)armed, glabrous. Fruits 4–5 (–8) mm long, 5–7 (–9) mm in diameter, depressed-cylindric, 6- to 8-costate, glabrous, purple-black when mature; pyrenes 45–50 in 6 or 8 radiating double-files, subequal, vertical. Figs. 3i, 6e, 20e.

Phenology. Flowering and fruiting throughout the year.

Distribution (Fig. 23). Eastern New Guinea and perhaps also the Geelvink Bay area of western New Guinea; occasional to frequent in primary or secondary montane forests, often in wet areas such as stream banks and seepage slopes, sometimes with *Nothofagus* or *Castanopsis*; 700–2150 m.

Vernacular names. *Komai* or *Homai* (Enga); *Tombrup* (Chimbu); *Toro* (Chimbu); *Aptun'no* (Kompiai); *Kainana* (Naho).

ADDITIONAL SPECIMENS EXAMINED. **Indonesia.** IRIAN JAYA [*Western New Guinea*]: Vogelkop Peninsula, Dalman, 45 km inland from Nabire, *Kanehira & Hatusima 12267* ♀ (A, BO). **Papua New Guinea.** [*Northeastern New Guinea.*] MADANG: Bundi patrol post, *ANU 10613* ♀ (CANB); vicinity of Budemu, S slopes Finisterre Range, *Pullen 6004* ♂ (A, BM, BRI, CANB, G, K, L, LAE); Moro, *NGF 21496* ♀ (A, BM, L, MO).—MOROBE: vicinity of Kikiepa, S slopes Finisterre Range, *NGF 11881* ♂ (A, BO, BRI, CANB, K, L, LAE), *NGF 11887* ♀ (A, BRI, CANB, K, LAE); Ogeramnang, *Clemens & Clemens 4747* ♂ (A, Z); Sattelberg, Lambanga, *Clemens & Clemens 7014* ♀ (A); Nomanenem camp, vicinity of Bulung River, *Clemens & Clemens 5206* ♀ (A); Kasanombe, rd to Momsalom, *NGF 37933* ♀ (A, E, K, L, LAE, M); gully N of Wagau airstrip, *NGF 17862* ♀ (A, BRI, CANB, E, K, L, LAE, SING); Wagau, *NGF 21572A* ♂ (BM, CANB, L, MO, SING); Mt. Shungol, ca. 5 mi SW of Wagau, *Hartley 12533* ♀ (A, BRI, C, CANB, G, K, L, LAE, RSA); track to Mt. Shungol, *NGF 45381* ♀ (A, BO, BRI, CANB, K, L, LAE, SING); Aseki-Menyamya rd, ca. 3 km from Aseki, *LAE 53940* ♀ (A, BO, CANB, K, L, LAE); vicinity of Haumga, *Craven & Schodde 1228* ♂ (A, BRI, CANB, K, L, LAE); near Kwaimengu, *Craven & Schodde 1429* ♀ (A, CANB, K, L, LAE).—WESTERN HIGHLANDS: 3 mi from Kopiago on Paga Hill Rd, *NGF 42067* ♀ (K, L, LAE); near Tibinini, 25 mi NW of Laiagam patrol post, *Robbins 3400* ♀ (CANB); Lufamanda, on bank of Lai River, *ANU 2719* ♀ (A, BRI, CANB, K, L, LAE); near Yaibos, Wabag, *ANU 2767* ♂ (A, BRI, CANB, K, L, LAE); Mt. Kum, *LAE 50215* ♂ (L, LAE); near Lyeiomanda, Timin River W of Kompiam, Sau Valley, *Robbins 2973* ♂ (CANB); Kompiai, *ANU 9610* ♂ (A, K, L, LAE); Koinambe, Jimi Valley, *Manner 18* ♀ (LAE); below Dagarunga Ridge, *NGF 38907* ♂ (A, BRI, CANB, K, L, LAE); vicinity of Wankl, ca. 5 km SE of Mt. Hagen station, *Hoogland & Pullen 5802* ♀ (A, BM, BRI, CANB, K, L, LAE); Kunsip, near Banz, Pinj River, *Robbins 567* ♂ (CANB, L, LAE); Uinba, Nona-Minj Divide, *Vink 16425* ♀ (A, BISH, C, CANB, K, L, LAE); Ongolth River, NE of Milep, *UPNG 4892* ♂ & ♀ (L, LAE, UPNG); Nondugl, *NGF 4387* ♀ (A, BO, BRI, CANB, K, L, LAE), *NGF 4950* ♂ (A, BISH, BM, BO, BRI, CANB, K, L, LAE, SING).—EASTERN HIGHLANDS: Warapuri and Warrangga Rivers, N of Nondugl, *NGF 18243* ♀ (A, BISH, BRI, CANB, K, L, LAE, SING); Sinasina, 1 mi W of Koge Mission, *Hide 92* ♀ (LAE); Purosa, Okapa area, *Brass 31846* ♀ (A, CANB, K, L, LAE); 5 mi S of Sassaura, *Hartley 12078* ♂ (A, CANB, K, L, LAE,

RSA); Arau, *Brass 32245* ♀ (A, CANB, K, L, LAE).—Not mapped: Southern Highlands: Hedemari area, *UPNG 2441* ♂ & ♀ (BISH, L, LAE, UPNG); Western Highlands: near Pina, ecological site no. 40, *ANU (Flenley) 2203* (CANB, L, LAE).

Timonius trichanthus is distinguishable from other species of subg. *Abbottia* in the shaggy-pubescent stipules and the leaf blades that are usually rounded at base. The species is here considered distinct from *T. cryptophlebus*, which is more densely pubescent and lacks leaf domatia, although that species, known only from the type, may prove to be a southward extension of *T. trichanthus*; should the two concepts be united, *Timonius cryptophlebus* would be the correct name. *Kanehira & Hatusima 12267*, the only collection from western New Guinea assigned here, is from an elevation of 400 m, and the flowers are noted as pink; the collection is therefore only tentatively placed in *T. trichanthus*.

27. Timonius trichocladus Merrill & Perry, J. Arnold Arb. 26: 239, fig. 3B. 1945.—TYPE: [INDONESIA. IRIAN JAYA.] Western New Guinea. Lake Habbema, 3225 m, Aug 1938, *Brass 9505* ♀ (holotype: A!; isotypes: BM! BO! BRI! L!).

Trees to 18 m tall and 30 cm dbh. Bark rough, brown to black, the wood white. Branchlets 2–4 mm wide toward apex, grayish, stramineous-strigillose to canescent, eventually glabrous. Stipules to 35 mm long and 8 mm wide, lanceolate to long-elliptic, acute, glabrous or puberulent on costa and margin abaxially, glabrous adaxially. Petioles 4–10 mm long, 1–1.5 mm wide at middle, semi- to subterete, glabrous to scattered-puberulent; leaf blades 5–9 cm long, 2–4 cm wide, elliptic to obovate, abruptly short-acuminate at apex, acute at base, coriaceous, glabrous except ciliate toward apex and base above, glabrous to scattered-sericeous toward margin and on costa beneath, the costa plane to canaliculate above, elevated beneath, the secondary nerves indistinct, the veinlets prominulous, paxillate; domatia pitlike, distributed along costa and on lamina beneath, puberulent. Staminate inflorescences 3- (5-)flowered, at anthesis ca. 2 cm long and 1.5 cm wide; peduncles 1–5 mm long, 1–1.5 mm wide, glabrous to strigillose; bracts absent, the bractlets absent or vaguely discernible beneath lateral flowers. Staminate flowers: calyx limb cupular, truncate to shallowly 4-toothed or sinuate, glabrous, the tube 1–1.5 mm long from base; corolla narrowly infundibular, white, the tube 6–8 mm long, 1.5–2.5 mm wide, glabrous outside, glabrous within, the lobes 4, 3–4 mm long, ca. 2 mm wide, ovate to deltate, acute, papillose, ciliate toward apex, the corolla limb 7–9 mm in diameter; stamens 4, the anthers ca. 4 mm long, 0.75 mm wide, linear-oblong, glabrous; pistillodium ca. 5 mm long, filiform, glabrous. Pistillate inflorescences 1-flowered; peduncles ca. 2 mm long (to 5 mm long in fruit), ca. 1 mm wide, glabrous; bracts absent. Pistillate flowers: calyx limb cupular, truncate, glabrous, the tube 0.5–1 mm long; corolla infundibular, white, the tube ca. 5 mm long, to 2 mm wide, glabrous outside, the lobes 4, 2.5–3 mm long, ca. 2 mm wide, ovate, acute, papillose, the corolla limb ca. 6 mm in diameter; staminodia not seen; hypanthium 4–5 mm long, 3–4 mm in diameter, ellipsoid, glabrous, the disc not seen, the style ca. 4.5 mm long, ca. 0.5 mm wide near middle, glabrous. Fruits 7–9 mm long, 6–7 mm in diameter, obovoid, glabrous, black when mature; pyrenes 15–25 in probably 4 radiating double-files, unequal, subvertical. Figs. 3h, 7c.

Phenology. Collected in flower and fruit from May to October.

Distribution (Fig. 23). Western New Guinea (Wissel Lakes and Lake Habbema); often locally abundant in dense, mossy forest or old secondary forest; 1800–3225 m.

Vernacular Name. *Ietowai* (Kapaukoe).

ADDITIONAL SPECIMENS EXAMINED. **Indonesia.** IRIAN JAYA [*Western New Guinea*]. Snow Mountains: Doglia, N of Kebo, Wissel Lakes, *BW 8789* ♂ & ♀ (L, LAE). Djajapura: 9 km NE of Lake Habbema, *Brass 10642* ♂ (A, BM, BO, BRI, L), *Brass & Versteegh 10486* ♂ (A, BM, BO, BRI, L), *Brass & Versteegh 11102* ♀ (A, BM, BO, BRI, L); Lake Habbema, *Brass 9504* ♂ (A, BM, BO, BRI, K, L, LAE).

Timonius trichocladus is distinguishable in having pubescent branchlets and obovoid fruits that are longer than wide and contain fewer than 30 subvertically oriented pyrenes. The available collections were all obtained from the Lake Habbema region of Western New Guinea in 1938, with an additional collection from Wissel Lakes, roughly 150 km to the west, made in 1960. Another collection, *Pullen 5605* ♂ (CANB, LAE), from forests near Siurane, Managalase area, Hydrographers Rang, Northern Province, Papua New Guinea, has branchlet indument characteristic of *T. trichocladus*, but is otherwise closer to *T. meridionalis*; that species is also known from southeastern New Guinea but has subglobose fruits enclosing obliquely arranged pyrenes.

28. Timonius wollastonii Wernham in H. Ridley, Trans. Linn. Soc. London, Bot. 9: 74. 1916.—TYPE: [INDONESIA. IRIAN JAYA.] Western New Guinea. Utakwa River to Mt. Carstensz, camp VIc (camp Pundok-Padang), 18 Feb 1913, *Boden Kloss* ♀ (lectotype, here designated: BM!).

Shrubs or trees to 6 m tall; nature of bark and wood unknown. Branchlets to 2 mm wide toward apex, pale to yellowish brown, finely canescent to scabridulous or with spreading reddish hairs to 0.5 mm long, often glabrate. Stipules to 35 mm long and 6 mm wide, lanceolate, acuminate, glabrous to canescent abaxially, glabrous adaxially except for colleters near base. Petioles 2–12 mm long, 0.5–1.5 mm wide at middle, semi- to subterete, puberulent to glabrous; leaf blades 1.5–5 (–8) cm long, 0.8–3 cm wide, rhombic-ovate, narrowed toward the apex but rounded to retuse at tip, acute to attenuate at base, thick-coriaceous, glabrous above, glabrous except often subciliate at apex beneath, the costa plane above, elevated beneath, the secondary nerves indistinguishable, the veinlets obscurely prominulous above, submerged beneath, paxillate; domatia pitlike, scattered over lamina and often also distributed along costa beneath, sometimes absent, glabrous. Staminate inflorescences 3- to 5- (7-)flowered, at anthesis to 1.5 cm long and 1 cm wide; peduncles 1–4 mm long, 1–1.5 mm wide, canescent; bracts absent. Staminate flowers: calyx limb cupular, truncate, glabrous, the tube 1–2 mm long from base; corolla narrowly infundibular, reddish with white lobes, the tube 5–7 mm long, ca. 0.5 mm wide, glabrous outside, glabrous within, the lobes 4, 1–1.5 mm long, ca. 0.5 mm wide, deltate, acute, glabrous, the corolla limb to 3.5 mm in diameter; stamens 4, the filaments short, the anthers 1.5–2.5 mm long, ca. 0.5 mm wide, linear-oblong, glabrous; pistillodium to 2.5 mm long, filiform, glabrous. Pistillate inflorescences 1-

flowered; peduncles 2–4 mm long, ca. 1 mm wide, glabrous; bracts absent. Pistillate flowers: calyx limb cupular, truncate, glabrous; corolla infundibular, reddish with white lobes, the tube 4–5 mm long, 1–1.5 mm wide, glabrous outside, glabrous within, the lobes 5, ca. 1.5 mm long in bud; staminodia 5, ca. 1.5 mm long, ca. 0.5 mm wide, linear, glabrous; hypanthium to 0.8 mm long, ca. 1 mm in diameter, obturbinate, glabrous to canescent, the disc glabrous, the style ca. 5 mm long, 5-armed, glabrous. Fruits 3.5–4 mm long, 2.5–5 mm in diameter, subglobose, 4- to 6-costate, glabrous, color of mature fruits unknown; pyrenes 12–20 in 4–6 radiating double-files, subequal, vertical. Figs. 3p, 6f, g.

Phenology. Collected in flower from October to February, in fruit in December.

Distribution (Fig. 23). Western central New Guinea; in primary forest; 1420–3200 m.

ADDITIONAL SPECIMENS EXAMINED. **Indonesia.** IRIAN JAYA [*Western New Guinea*]. Djajapura: near Mamberamo River, ridge to Doormantop, *Lam 1485* ♂ & ♀ (A, BO, L), *Lam 1905* ♀ (BO, L); Doormantop, *Lam 1784* ♂ (BO, L). Maoke Mountains [Snow Mountains]: Utakwa River to Mt. Carstensz, camp VIc, *Boden Kloss s.n.* ♂ (BM); Bijenkorf bivouac, *Pulle 825* ♀ (BO, L); crest of Hellwig Mountains, *Pulle 888* ♂ (BO, L). Without further locality, *von Römer 1242* ♂ (BO).

Timonius wollastonii resembles *T. avenis* but occurs at higher elevations and is characterized by bluntly acute, rounded, or retuse leaf blades, and fruits enclosing 20 or fewer pyrenes. It is thus far known only from western New Guinea, but a Papuan collection, *van Royen 10985*, from 2670 m, also has thick-coriaceous and retuse leaf blades, but otherwise resembles *T. meridionalis* (see that species, no. 15). The type collection of *T. wollastonii* includes staminate and pistillate material; Wernham's brief description seems based upon the pistillate element, which is here chosen as the lectotype.

29. Timonius zuckianus S. Darwin, J. Arnold Arb. 64: 611. 1983.—TYPE: PAPUA NEW GUINEA. Milne Bay: Normanby Island, Mt. Pabinama, 820 m, 2 May 1956, *Brass 25653* ♀ (holotype: A!; isotypes: CANB! K! L! LAE! US!).

Shrubs or trees to 20 m tall. Bark rough, dark brown, the wood white to pale straw-colored. Branchlets 2–5 mm wide toward apex, dark purple-black, somewhat glaucous, glabrous. Stipules to 130 mm long and 15 mm wide, lanceolate, acuminate, glabrous abaxially, glabrous adaxially except at very base among colleters. Petioles 12–55 mm long, ca. 2 mm wide at middle, semi- to subterete, glabrous; leaf blades 7–13 cm long, 3–7.5 cm wide, elliptic (usually broadly so), abruptly and sharply acuminate-mucronate at apex with a mucro 2–5 mm long, obtuse to acute to attenuate at base, thick-coriaceous, glabrous and lustrous above, glabrous beneath, the costa plane to canaliculate above, elevated beneath, the secondary nerves indistinct, the veinlets prominulous above, plane to prominulous beneath, paxillate; domatia pitlike, distributed over lamina beneath, glabrous to puberulent. Staminate inflorescences 5- to 7-flowered, at anthesis to 3 cm long and 3 cm wide; peduncles 6–10 mm long, to 3 mm wide, compressed, glabrous; bracts absent. Staminate flowers: calyx limb cupular, truncate, glabrous, the tube 2.5–4.5 mm long from base; corolla narrowly infundibular, white, the tube 8–12 mm long, ca. 1.5 mm wide, glabrous outside, glabrous within, the lobes 4, 6–8 mm long, 3–4 mm wide, ovate, acute to somewhat erosulous, papillose, the corolla limb 12–18 mm in diameter; stamens 4, the filaments short, the anthers 5–6 mm long, ca. 0.5 mm wide, linear, glabrous; pistillodium to 3 mm long, filiform, glabrous. Pistillate inflorescences 2- to 4-flowered, at anthesis to 3.5 cm long and 2 cm wide; peduncles 8–18 mm long, to 3 mm wide, compressed, glabrous; bracts absent. Pistillate flowers: calyx limb cupular, truncate to undulate, glabrous, the tube 0.5–1 mm long; corolla hypocrateriform, white, the tube 5–7 mm long, ca. 1.5 mm wide, glabrous outside, glabrous within, the lobes 6–8, to 5 mm long, 2 mm wide, ovate, narrowly acute to acuminate, papillose, the corolla limb to 12 mm in diameter; staminodia 6 (or perhaps also 8?), ca. 3 mm long, ca. 0.5 mm wide, linear-sagittate, glabrous; hypanthium 3–5 mm long, 3–5 mm in diameter, subglobose, glabrous, the disc glabrous, the style 6–10 mm long, ca. 0.5 mm in diameter near middle, irregularly 6-armed, glabrous. Fruits 4–5 mm long, 6–7 mm in diameter, short-cylindrical, 6-ribbed, glabrous, black when mature; pyrenes 35–45 in 6 radiating double-files, subequal, vertical. Figs. 3b, 20f, g.

Phenology. Probably flowering and fruiting throughout the year.

Distribution (Fig. 16). Islands of Milne Bay Province of Papua New Guinea; common in lower montane forests and stunted mossy forests of ridge crests; 700–1700 m.

ADDITIONAL SPECIMENS EXAMINED. **Papua New Guinea.** [*Papua.*] MILNE BAY: Goodenough Island, E slopes, *Brass 24525* ♀ (A, CANB, L, LAE); Fergusson Island, Mt. Kilkerran, E slopes (Woiatabu), *LAE 71045* ♂ (A, BISH, E, K, L, LAE, M); Normanby Island, Mt. Rumabubu, summit, *LAE 67868* (L, LAE); Tagula Island, Mt. Riu, W slopes, *Brass 27839* ♀ (A, L, LAE), Mt. Emuwa, *LAE 74520* ♀ (A, BRI, CANB, K, L, LAE, QRS); Rossel Island, Kwa mountain above Abilete, *NGF 27077* ♀ (A, BRI, CANB, K, L, LAE), Mt. Rossel, summit, *Brass 28458* ♀ (A, BO, K, L, LAE), Mt. Rossel, S slopes, *Brass 28389* ♂ (A, BO, L, LAE).

Timonius zuckianus is unusual in its exceptionally large flowers and prominently acuminate leaf apices. It resembles *T. longifolius*, from the Sepik region of New Guinea, in its cymose pistillate inflorescences, but the larger fruits, glabrous inflorescences, and elliptic rather than oblanceolate leaves suggest *T. singularis*, from Australia. One collection from the upper Kumusi Valley of Papua, northeast of Emo, Northern Province [*Paijmans 156* ♂ (CANB)], is probably assignable to *T. zuckianus*, but has somewhat smaller flowers and more prominent leaf domatia, approaching *T. meridionalis*, which occurs in the same region.

ACKNOWLEDGMENTS

Sincere appreciation is expressed to the administrators of the following institutions, who kindly made herbarium collections available for study: A, B, BISH, BM, BO, BR, BRI, C, CANB, DAV, E, G, GH, K, L, LAE, M, MASS, MEL, MO, NO, NY, P, PNH, QRS, RSA, S, SING, SUVA, U, UPNG, US, W, Z. Michael Galore and his staff at LAE provided invaluable assistance with field work in Papua New Guinea. Paul Kores assisted with photomicroscopy, and William Schley, Jr., made anatomical preparations. I thank Mia Molvray for providing translations of critical papers in German and Dutch, and for gathering herbarium specimen data at MEL. Christopher Puttock very kindly reviewed locality data for Australian collections and made other corrections to the final manuscript draft. To Christian Puff and Elmar Robbrecht I am especially indebted for advice that significantly improved the manuscript.

LITERATURE CITED

Bailey, F. M. 1900. *The Queensland flora*. Brisbane: H. J. Diddams. [Rubiaceae, 742–783.]

Baillon, H. E. 1880. *Histoire des plantes*, vol. 7. Paris, London, Leipzig: L. Hachette et Cie. [Rubiaceae, 257–503.]

Bentham, G. 1867. *Flora australiensis*. Vol. 3. London: Lovell Reeve & Co. [Rubiaceae, 399–477.]

Chaw, S.-M., and S. P. Darwin. 1992. A systematic study of the paleotropical genus *Antirhea* (Rubiaceae: Guettardeae). Tulane Studies Zool. Bot. 28: 25–118.

Darwin, S. P. 1983. New species of *Timonius* (Rubiaceae) from Papuasia. J. Arnold Arb. 64: 611–618.

———. 1993. A revision of *Timonius* subg. *Timonius* (Rubiaceae: Guettardeae). Allertonia 7: 1–39.

Darwin, S. P., and Chaw, S.-M. 1990. Bobea. In *Manual of the flowering plants of Hawaii*, ed. W. L. Wagner, D. R. Herbst, and S. H. Sohmer, 2: 1114–1118. Honolulu: University of Hawaii Press; Bishop Museum Press.

Elmer, A. D. E. 1911. New and noteworthy Rubiaceae. Leafl. Philipp. Bot. 3(54): 971–1046.

Erdtman, G. 1960. The acetolysis method–a revised description. Svensk. Bot. Tidskr. 54: 561–564.

Foster, A. S. 1950. Morphology and venation of the leaf in *Quiina acutangula* Ducke. Amer. J. Bot. 37: 159–171.

———. 1952. Foliar venation in angiosperms from an ontogenetic standpoint. Amer. J. Bot. 39: 752–766.

Gray, A. 1858. Notes upon some Rubiaceae. Proc. Amer. Acad. Arts 4: 33–50.

Jacobs, M. 1966. On domatia – the viewpoints and some facts. Proc. Kon. Ned. Akad. Wetens., Ser. C, 69: 275–316.

Luo, X.-R., Y.-H. Liang, and C. T. Chen. 1993. Pollen morphology of Chinese Rubiaceae (abstract). International Conference on the Systematics of the Rubiaceae, Missouri Botanical Garden, p. 13.

Martinello, M. 1992. Fruchtknoten und Frucht von *Timonius* DC. (Rubiaceae-Antirheoideae-Guettardeae) und deren Entwicklung, unter besonderer Berücksichtigung von *T. flavescens* (Jack) Baker. Dipl.–Arbeit (M.S. thesis), Form.–Naturw. Fac., Universität Wien.

Melville, R. 1976. The terminology of leaf architecture. Taxon 25: 549–561.

Merrill, E. D. 1909. New or noteworthy Philippine Plants, VII. Philipp. J. Sci. 4: 247–330.

———. 1923. *An enumeration of Philippine flowering plants*. Manila. [Rubiaceae, 3: 492–576.]

———. 1925. Additions to our knowledge of the Philippine flora, I. Philipp. J. Sci. 26: 447–496.

Merrill, E. D., and L. M. Perry. 1945. Plantae Papuanae Archboldianae, XVI. J. Arnold Arb. 26: 229–266.

Metcalfe, C. R., and L. Chalk. 1950. *Anatomy of the dicotyledons*. Oxford: Oxford University Press. [Rubiaceae, 759–776.]

———. 1979. *Anatomy of the dicotyledons*, ed. 2, vol. 1. Oxford: Oxford University Press.

Moore, S. 1923. *Timonius cryptophlebus*. In "A contribution to our knowledge of the flora of Papua" by C. T. White, Proc. Roy. Soc. Queensland 34: 5–65.

Mueller, F. von. 1861. *Fragmenta phytographiae australiae*, vol. 2. Melbourne: Government of Victoria. [Rubiaceae, 132–135.]

———. 1875. *Fragmenta phytographiae australiae*, vol. 9. Melbourne: Government of Victoria. [Rubiaceae, 179–188.]

Nicolson, D. A. 1979. *Neolaugeria*, a new name for *Terebraria* (Rubiaceae) of the West Indies. Brittonia 31: 119–124.

O'Brien, T. P., and I. von Teichman. 1974. Autoclaving as an aid in the clearing of plant specimens. Stain Tech. 49: 175–176.

Parham, B. E. V. 1972. Plants of Samoa. New Zealand Dept. Sci. Indust. Res. Inform. Ser. 85: 1–163.

Puff, C. 1986. A biosystematic study of the African and Madagascan Rubiaceae–Anthospermeae. Pl. Syst. Evol., suppl. 3: 1–535.

Robbrecht, E. 1988. Tropical woody Rubiaceae. Opera Bot. Belg. 1: 1–271.

Schumann, K. 1891. Rubiaceae. In *Die natürlichen Pflanzenfamilien*, ed. A. Engler and K. Prantl, IV. 4: 1–156. Leipzig: Wilhelm Engelmann.

Smith, A. C., and S. P. Darwin. 1988. Rubiaceae. In *Flora vitiensis nova* by A. C. Smith, 4: 143–376.

Smith, L. S. 1957. New species of and notes on Queensland plants–II. Proc. Roy. Soc. Queensland 68: 43–50.

Takhtajan, A. 1989. *Floristic regions of the world*. Berkeley: University of California Press.

Ter Welle, B. J. H., A. A. Loureiro, P. L. B. Lisboa, and J. Koek–Noorman. 1983. Systematic wood anatomy of the tribe Guettardeae (Rubiaceae). Bot. J. Linn. Soc. 87: 13–28.

Valeton, T. 1909. Beiträge zur Kenntniss der Gattung Timonius. Bull. Dép. Agric. Indes Néerl. 26: 1–61.

———. 1927. Die Rubiaceae von Papuasien. Zweiter Teil: Coffeoideae. Bot. Jahrb. Syst. 61: 32–163.

van Steenis, C. G. G. J. 1981. *Rheophytes of the world.* Alphen a/d Rhijn: Sijthoff & Noordhoff.

Verdcourt, B. 1958. Remarks on the classification of the Rubiaceae. Bull. Jard. Bot. Etat Bruxelles 28: 209–281.

Wernham, H. F. 1916. Dr. H. O. Forbes's New Guinea Rubiaceae–II. J. Bot. 56: 129–135.

Wong, K.-M. 1988. The *Antirheoideae* (Rubiaceae) of the Malay Peninsula. Kew Bull. 43: 491–518.

Zollinger, H. 1857. Observationes botanicae novae. Natuurk. Tijdschr. Ned.-Indië 14: 144–176.

NUMERICAL LIST OF SPECIES

1a. T. affinis var. affinis
1b. T. affinis var. samoensis
1c. T. affinis var. sapotifolius
2. T. amungwiwanensis
3. T. appendiculatus
4a. T. avenis var. avenis
4b. T. avenis var. magnifructus
4c. T. avenis var. pubipetalus
4d. T. avenis var. vogelkopensis
5. T. belensis
6. T. bismarckensis
7. T. bougainvillensis
8. T. carstensensis
9. T. cryptophlebus
10. T. dolichophyllus
11. T. heptamerus
12. T. lamii
13. T. longifolius
14a. T. longitubus var. longitubus
14b. T. longitubus var. pubescens
15. T. meridionalis
16. T. modestus
17. T. neobritannicus
18. T. oktediensis
19. T. pachyphyllus
20. T. pseudaffinis
21a. T. pubistipulus var. pubistipulus
21b. T. pubistipulus var. pubescens
22. T. scabriflorus
23. T. secundiflorus
24. T. singularis
25. T. subavenis
26. T. trichanthus
27. T. trichocladus
28. T. wollastonii
29. T. zuckianus

INDEX TO NUMBERED COLLECTIONS EXAMINED

The numbers in parentheses refer to the corresponding species in the text and in the Numerical List of Species presented above.

Aët & Idjan 314 (4b).

Allison, A. NG-161 (21b); NG-171 (21b); NG-195 (21b).

ANU 762 (21a); 2203 (26); 2323 (5); 2719 (26); 2767 (26); 2805 (21b); 7400 (5); 9610 (26); 10613 (26); 10623 (5); 20326 (21a); 20353 (21a).

bb 34694 (4a).

Blake, S. T. 15220 (24).

Bowers 376 (5); 528 (5).

Brass, L. 2295 (24); 3244 (14a); 3316 (14a); 4077 (21a); 4705 (2); 5046 (21a); 6748 (16); 6804 (10); 7076 (16); 9504 (27); 9505 (27); 10642 (27); 11527 (5); 12398 (10); 12878 (4c); 12913 (23); 13310 (23); 13628 (23); 13687 (4c); 20338 (24); 22304 (21a); 22702 (21a); 22874 (21a); 24525 (29); 24773 (15); 24796 (15); 25082 (15); 25653 (29); 27460 (15); 27532 (15); 27839 (29); 28389 (29); 28449 (15); 28458 (29); 28630 (15); 29559 (21b); 29561 (21b); 30219 (5); 30346 (5); 30619 (5); 30837 (5); 31056 (5); 31093 (5), 31846 (26); 32230 (22); 32245 (26).

Brass & Versteegh 10486 (27); 11102 (27); 11110 (5); 12592 (4c).

BSIP 869 (20); 907 (14a); 1498 (20); 1696 (1a); 1787 (1a); 1884 (14a); 1902 (20); 2021 (14a); 2285 (14a); 2387 (20); 2536 (20); 2690 (20); 3907 (20); 4032 (14a); 5067 (14a); 5103 (14a); 5192 (20); 5349 (20); 5367 (14a); 5393 (14a); 5862 (14a); 6006 (14a); 6121 (14a); 6424 (14a); 6678 (14a); 6770 (14a); 7123 (14a); 7150 (14a); 7519 (20); 8275 (14a); 8526 (20); 9476 (20); 10107 (14a); 11193 (14a); 11474 (20); 11654 (20); 12056 (20); 12323 (14a); 13042 (14a); 13104 (14a); 13163 (14a); 13670 (14a); 13864 (14a); 13963 (14a); 14290 (14a);

14443 (14a); 15191 (20); 15627 (14a); 15888 (14a); 15952 (14a); 15982 (14a); 16010 (14a); 16249 (14a); 16258 (20); 16528 (14a); 16609 (14a); 17019 (1a); 17525 (14a); 18423 (14a); 18604 (14a); 18754 (14a).
Bulmer, R. 134 (5).
BW 4975 (4d); 8101 (25); 8789 (27); 10275 (4a); 11462 (12); 11936 (4d); 12377 (16); 14130 (4b); 15269 (4d); 15353 (4d).
Carr, C. E. 12522 (21a); 13095 (21a); 13343 (21a); 13819 (21a); 14607 (21a); 15042 (21a); 15856 (21a); 16453 (21a); 16478 (21a).
Christophersen, E. 3246 (1b).
Clemens, J., & Clemens, M. 3568bis (22); 3586a (22); 4143 (22); 4649 (21b); 4747 (26); 5206 (26); 5758 (21a); 7014 (26); 11158B (26).
Coode, M. 3885 (22).
Corlett, R. 25162 (2); 25164 (2).
Cox, P. A. 192 (1b).
Craven & Schodde 207 (14a); 1228 (26); 1291 (21b); 1429 (26).
Crome, F. 884 (24).
Darbyshire, P. J. 455 (23).
Darbyshire & Hoogland 8355 (13).
Dockrill, A. W. 33 (24); 1434 (24).
Docters van Leeuwen, W. M. 10141 (25); 10172 (25).
Elmer, A. D. E. 15639 (3); 16636 (3).
Eyma, P. J. 2552 (4a).
Fallen, M. 550 (21a).
Fiji Dept. Agriculture 2988 (1a); 8509 (1a); 12719 (1a); 13712 (1a); 15751 (1a); 17018 (1c).
Forbes, H. O. 770 (21a).
Gillespie, J. W. 2620 (1a); 2890 (1a); 3106 (1a); 3552 (1a); 3755.5 (1a); 4292 (1a); 4342 (1a); 4468 (1c); 4512 (1a); 4613 (1a); 4817 (1a).
Gray, B. 4564 (24).
Greenwood, W. 873 (1a); 873A (1a).
Hartley, T. G. 11725 (21b); 11840 (21b); 12078 (26); 12430 (22); 12533 (26); 12716 (21a); 13222 (23); 13687 (5).
Hays, T. E. 269 (22); 363 (22).
Hide, R. L. 92 (26); 309 (21a); 486 (21a).
Hoogland, R. D. 9600 (21a).
Hoogland & Craven 9341 (21b); 10663 (13); 10747 (13); 10950 (23); 10986 (23).
Hoogland & Pullen 5802 (26).
Hoogland & Schodde 7089 (21a).
Hyland, B. 3583 (24); 5235 (24); 7020 (24); 7621 (24); 11440 (24); 13596 (24).
Hyndman, D. 300 (5).
Irvine, A. K. 819 (24); 2284 (24).
Jacobs, M. 7934 (3); 8859 (21b); 8894 (5); 9230 (16).
Jermy, A. C. 5025 (21b).
Jessup & Clarkson 290 (24).
Kajewski, S. F. 623 (1a); 1191 (24); 1677 (7); 1750 (7); 2009 (14a); 2049 (7); 2541 (14a).
Kalkman, C. 4337 (10); 4384 (10); 4731 (5); 5040 (5).
Kanehira & Hatusima 11667 (16); 12267 (26); 12268 (16); 13141 (4d); 13608 (4b).
Kostermans, A. J. G. H. 1098 (16); 1107 (16); 2220 (4b); 2265 (4b); 5229 (4a); 5711 (4a).
Kostermans & Soegeng 767 (12).
LAE 50215 (26); 50320 (21a); 51091 (5); 51169 (6); 53940 (26); 54117 (21a); 54178 (21a); 54772 (21b); 55321 (21a); 59483 (21b); 59583 (4c); 61164 (4a); 61972 (21a); 63162 (5); 65579 (6); 66113 (17); 66351 (5); 67013 (5); 67868 (29); 71045 (29); 73981 (25); 74437 (21a); 74520 (29); 74603 (15); 75226 (21a); 76014 (17); 77669 (21b); 77670 (21b).
Lam, H. J. 1485 (28); 1572 (12); 1784 (28); 1905 (28).
Ledermann, C. L. 8212 (13); 8213 (13); 8426 (23); 8479 (23); 8793 (25); 8899 (23); 9853 (23); 9904 (23); 9945 (23); 10292 (23); 12678 (4c); 12870 (4c).
Mangen 356 (5).
Manner, H. 18 (26).
Melville & Parham 7108 (1a).
Meyer Drees, E. 499 (25); 607 (25).
Moi, W. 107 (2).
Moriarty, V. K. 2028 (24).
NGF 4387 (26); 4950 (26); 6751 (5); 11285 (5); 11375 (5); 11727 (5); 11881 (26); 11887 (26); 13489 (21a); 13908 (21b); 17862 (26); 18173 (21b); 18243 (26); 19948 (21b); 19959 (21b); 20308 (21a); 21063 (21b); 21272 (5); 21496 (26); 21572A (26); 21945 (17); 24225 (14b); 24614 (5); 25119 (5); 25188 (5); 26517 (17); 26763 (6); 26986 (5); 27077 (29); 27241 (17); 27355 (6); 27372 (6); 27552 (21a); 28497 (5); 28897 (21a); 29135 (21b); 29977 (15); 30650 (14a); 31299 (7); 31382 (7); 31572 (7); 31978 (4a); 32144 (21a); 32181 (21a); 32502 (21a); 33115 (10); 33179 (10); 33417 (4a); 34411 (17); 34629 (21a); 35512 (21a); 36847 (2); 37434 (21a); 37933 (26); 38024 (6); 38907 (26); 40730 (5); 41515 (4c); 41516 (4c); 41766 (4c); 42067 (26); 42562 (23); 42793 (18); 42807 (4c); 43844 (16); 45381 (26); 49482 (6).
Paijmans, K. 156 (29); 1090 (21a); 1257 (5); 1568 (21b).
Parks, H. E. 20891 (1a).
Philipp. Bur. Sci. 1505 (3); 2164 (3); 23330 (3); 29591 (3); 29778 (3); 40794 (19); 42277 (19).
Philipp. Nat. Herb. 10583 (3); 21621 (3).

INDEX TO SCIENTIFIC NAMES

Accepted names are in roman type; the main entry for each is in **boldface**. Synonyms are in *italics*.